WEATHERBEATEN

Dedication

To Colin McMillan, Graham McMillan and Allen Vamplew,
whose interest in sport inspired ours.

Acknowledgements

We should like to thank Ian Beesley, and other suppliers of photographs, the Carnegie Trust, Clare McMillan, Ailsa Vamplew and the staff of the Sports Turf Research Institute at Bingley for their assistance. We are particularly indebted to John Burnett for inadvertently launching us on this project.

WEATHER BEATEN

BEATEN

Sport in the British Climate

Joyce Kay and Wray Vamplew

MAINSTREAM
PUBLISHING
EDINBURGH AND LONDON

First published in Great Britain in 2002 by
MAINSTREAM PUBLISHING (EDINBURGH) LTD
7 Albany Street
Edinburgh EH1 3UG

ISBN 1 84018 602 X

A catalogue record for this book is available from the British Library

Typeset in Allise and Garamond

Printed in Great Britain by
Creative Print and Design Wales

Contents

Grounds for Divorce?

There may be Trouble Ahead

Always Take the Weather with You

Saturation Point

Foreword

My formative years were spent in the Pennine region of West Yorkshire and there is no doubt that the severe winter of 1963 served as the catalyst for my career in meteorology. Joyce and Wray have devoted a whole chapter to that infamous period of bad weather but from a personal point of view the first snowfall on Boxing Day resulted in three months of torture for a sport-loving schoolboy who had to endure a frozen playground and risk life and limb every time I walked to school.

The government of the day was so concerned about a repeat of severe disruption to the sporting calendar and infrastructure of the country that they instructed the Meteorological Office to produce 30-day forecasts on a regular basis. Of course it was an impossible task but refusal was not an option. Forty years later it is still not possible, a few months ahead, to forecast a spell of severe winter weather when the football season could hibernate for a month.

Everyone has a story to tell about how their favourite pastime has been ruined by inclement weather and this book relates the multitude of well-known crises that have affected the sporting calendar over the last century. On almost every page you can massage the memory banks and in many cases acknowledge that you were there or at least relate your own stories to particular noteworthy events. This is an anthology through which you can relive the past and which proves that the weather and sport are inextricably linked, often detrimentally but occasionally to the enhancement of certain sporting events.

Inevitably the Great Storm of 15–16 October 1987 deserves a mention in Chapter 2, with the infamous comments from my former colleague Michael Fish who said there was no need to worry, there wouldn't be a hurricane. It was particularly disappointing for me that the severity of the storm was played down as it approached because I had been responsible for highlighting the potential for some very windy weather when presenting the Farming Forecast on BBC1 the previous Sunday. Only the evening before, another colleague, Bill Giles, mentioned that it was becoming 'breezy up the Channel', surely the understatement of the decade. However, it is worth noting that the seldom-mentioned Burns' Day Storm of 1990 caused a greater death toll and affected a larger part of the UK.

It remains to be seen how global warming will impact on sport in the British climate in the next century. If present research is to be believed, and recent changes have to be acknowledged, then we can expect more severe storm events but also a warmer world with droughts being more common for south-east Britain. Clearly there will be plenty of ammunition for a follow-up to this volume but in the meantime I can only delight in the efforts of Joyce and Wray in producing this excellent book that will be enjoyed by readers young and old, whether sporting fans or not.

John Kettley

Preface

The British are obsessed with the weather. This is not surprising: are two days ever the same? When have the seasons obeyed their rightful slot in the calendar? Many Brits are equally preoccupied with sport. Bring the two together and you end up with questions such as: under what conditions did goalkeeper Pat Jennings score for Spurs? How often has rain stopped play at Old Trafford? What caused the 1962–63 rugby league season to finish on 1 June? Why doesn't Wimbledon have a retractable roof to prevent fiascos like the rain-delayed semi-final between Tim Henman and Goran Ivanisevic in 2001?

This book answers these and other questions, and takes a look at how fog, rain, snow, frost, storm – and occasionally sunshine – have affected sport in Britain. It is not a technical book about weather, only about how it affects sport. It provides a variety of examples from many sports and most weather conditions, mainly from the last 50 years, but does not claim to be comprehensive. It couldn't be – there is too much of both around! If your favourite anecdote, gut-wrenching hard-luck story or definitive illustration has not been mentioned, we plead lack of space.

No one in England will forget that glorious day in 1966 when the national football team won the World Cup. (Many in Scotland, however, have tried!) But memory can be faulty; the sun may have shone gloriously when Bobby Moore lifted the Cup but, according to contemporary press reports, the afternoon had been squally with several heavy showers. Although we have used newspapers extensively in our research, this has tended to confirm that there is no such thing as 'truth', only opinion and interpretation!

A few technical points need further explanation. The words climate and weather are often used interchangeably, but we have taken climate to mean the backdrop against which the daily weather occurs. Wet springs, dry summers, late autumns and sometimes harsh winters are all part of the British climate (or dry springs, wet summers, cold autumns and mild winters – this is Britain, after all!). The weather is what we find when we pull the curtains each morning.

Whatever the conditions, they are often expressed statistically and we are aware that Britain in the twenty-first century has a problem with measurement. Half the population feel more comfortable with feet and inches, stones and pounds and Fahrenheit temperatures and less comfortable with their metric equivalents. In sport, horse-racing measures its races in furlongs and its jockeys in pounds and golfers still refer to yardage charts while yachts and yachtsmen deal in feet, fathoms and knots. In this book the measurements used are metric – rainfall in millimetres, wind speeds in kilometres per hour and temperatures in Celsius – except where names of events and historical circumstances warrant otherwise. However, we have provided a conversion chart to help make sense of temperatures.

The importance of the weather to sport has long been recognised. Even in the 1930s, sportswriters saw it as one of the chief factors in deciding the result of sporting events. Seems strange that few have chosen to examine the topic more closely – till now!

UNDER THE WEATHER

1

Weatherbeaten?

We all have a personal memory of how weather affected a sports event dear to us. We may have been there in person, standing on the terraces in the pouring rain or sitting in the sun with a warm beer. We may have been watching in the comfort of the pub or the living room as some poor athletes slogged it out on a muddy pitch or put in the final lung-bursting effort on a baking hot day. We may have listened to the radio in despair as our team succumbed to the rigours of some foreign climate. We may even have taken part in some glorious or ghastly contest in which the weather enhanced or wrecked the occasion.

Weather is a chancy affair for everyone connected with sport: officials and administrators, spectators, the media, the forecasters and, most of all, the competitors in every outdoor sport, at any level. It's a bit like the lottery of which referee is assigned to a match. Will it be easy-going and benign – a fatherly talking to, no yellow cards? Will it be blustery and harsh – a sharp-tongued, red card specialist? Or will it be diabolical and unfair – awarding a penalty for a blatant dive in stoppage time? The only certainty is that the behaviour of match officials and weather are beyond our control.

It is pouring with rain as we write this and a brisk wind is scattering the last of the spring blossom. Fortunately, it is a Monday morning. If it were a Saturday or Sunday, thousands of sports fans and players throughout the country would be bracing themselves for a grim afternoon of trench warfare on the muddy sports fields of Britain. It would be an outing for the more robust footballing skills, the resolute defence, the direct approach: no place for the fancy dan! Not much chance of flowing rugby or smooth, passing football on a day when any lofted ball would be held up in the swirls and eddies of the stadium or the sudden gusts of a windswept park, or bogged down in the end-of-season goalmouth swamp. A day, perhaps, when the opposition might be tempted to try out the tactics allegedly used against Stanley Matthews in an FA Cup fifth round tie in 1954. Port Vale were said to have spent large parts of the game forcing Matthews away from the drier areas on the wing to the muddier midfield. The tactic worked as they prevented him from scoring and Blackpool lost 2–0.

Weather that is unhelpful to one contestant will often benefit another. While one cricket team is warily eyeing the advancing dark clouds, the other is digging in and praying for rain . . . soon! When Pakistan scored 608 in their first innings against England at Edgbaston in 1971, the home side had to follow on for the first time against their Asian opponents. At the end of the fourth day they had still not reached the Pakistan total and had only seven wickets remaining. However, the rain gods were not smiling on the visitors. Fewer than 15 overs were possible on the final day; Pakistan were thwarted, the match was drawn and England went on to win the series. The weather has been responsible for many similar lost opportunities and near misses.

It is also one of the few forces that can disrupt sport. Although royal deaths, such as that of Princess Diana, have brought the Saturday sports programme to a halt, and wartime has greatly modified and shrunk the amount on offer, the show normally goes on – except when the weather intervenes. The following pages are crammed with examples of how snow, frost, rain and fog have wiped out single events or entire leagues, sometimes for weeks on end. A familiar sign of winter, the Pools Panel resulted from an unplanned shut-down in the football season; nearly 40 years on, a planned suspension of league programmes is advocated by someone, somewhere in Britain whenever wintry weather disrupts the fixture list. Bookmakers regularly offer odds on White Christmases but will also quote for rain or sunstroke at Wimbledon or the date and place where turf racing will resume after a lengthy weather break. In 1963, with no racing for months and no all-weather tracks, some bright spark suggested a horse-race panel in imitation of the new football body. Bookies found the idea hilarious, pointing out that they made a living because the 'experts' were so often wrong!

It is not just headline-grabbing, big-time sport which is at the mercy of the weather. The shinty season in Scotland is affected every bit as much by winter rain and snow as the more ubiquitous football and rugby seasons. Racing pigeons and their handlers are as upset by fog or strong winds as professional golfers. Anglers will have welcomed the mild, dry April of 2002 as much as first-class cricketers. If we have not found space to mention your particular sport, it is not because it is immune from weather hazards. No outdoor sport is ever free of their influence; no amount of planning, expenditure and expertise can protect an event from the climate, from the lucrative Seve Ballesteros Trophy at Druids Glen, Ireland in 2002 to the humblest amateur cup tie.

Modern gadgetry such as penetrometers (to measure the going on racecourses) and light meters (to gauge the visibility at cricket grounds) often simply confirm the weather is as bad as we think. Technology and twenty-first century expectations have sometimes even added to the problems of sport. Wraparound stadiums have played havoc with pitches and television's demands for attractive green turf have led to much slipping and sliding on the greasy surfaces now commonplace at major football grounds. A whole section of this book is devoted to the groundsman's age-old problems – and some new ones – as well as advances in the use of artificial grass.

When we set out to write this book, there were neat chapters labelled Winters and Summers. Then we discovered something which should have been obvious: winter can still be with us in May while summer can make a brief appearance in April. How else could administrators justify having the end of the football season in the same month as the Chelsea Flower Show or the start of the cricket season before the arrival of the cuckoo? As easterly winds sometimes scythe across the bleak exposed heath at Newmarket, and frosts wipe out many a darling bud, so the FA Cup final will often be played on balmy days. Sevens tournaments end the rugby programme, often in bright spring sunshine, while first-class cricketers frequently have to endure hours on grey, sleet-strewn fields, fortified only by the occasional hot tea or soup.

Britain's weather is, to put it nicely, capricious. As we finish this chapter, the morning rain has passed, the sun has appeared and the wind picked up, sending clouds scudding across the sky. Heavy rain and hailstones have been reported in Lancashire – how might that affect the possible Premiership decider tonight when Arsenal take on Bolton at the Reebok Stadium? (It didn't!) How is the weather in Germany where Manchester United are preparing for the crucial second leg of their Champions League semi-final against Bayer Leverkusen? (They lost anyway!) Britain is not on its own, however. Other countries also have their share of unseasonal, unhelpful and downright awful weather, as discussed towards the end of this book.

Favourite weather memories? You'll find a few scattered throughout this book, but one of us particularly recalls a beautiful summer dawn and a sunny early morning journey to London with the prospect of a fine day at Lord's. As the train crept into north London, a few spots of rain appeared on the windows. Within half an hour, there was a steady downpour in a cool breeze. Joining the crowds thronging St John's Wood for England v Australia, armed with their picnic baskets and panamas, there was optimism in the air – surely it would be all right? Four hours later, it patently wasn't. The rain had eased, stopped . . . and started again. Occasional announcements were made, umpires and players made brief appearances and retreated again to the warmth of the dressing-rooms, while the poor sods in the stands became progressively colder, damper and more disgruntled. Having fought our way into the Lord's shop to pass some time and exhausted the delights of the catering facilities – too cold for beer – there was little choice but to pack up and go home before hypothermia set in. Such is sport . . . and the weather!

2

Blowing in the Wind

Illiam Gilbert Grace, the great cricketer, shamateur and cheat, reputedly once refused to leave the crease as he claimed that the wind was responsible for removing the bails from his stumps, not the ball. W.G. may have been able to intimidate umpires when he played but, in modern cricket, special heavy-duty bails are used when the wind is strong so such chicanery has little chance of success. On Britain's breezy playing fields, wind is a fact of life for sportsmen and women. Its effects have to be contended with on a regular basis, hopefully by means less unsporting than those of the good doctor.

Day in, day out, when the wind blows, decisions have to be made. If you win the toss do you elect to play with the breeze or against it? Which bowler does a cricket captain ask to trundle into the wind and to which does he give the assistance of the wind? Should normal rugby kicking tactics be abandoned in the teeth of a gale blowing down the pitch? Or up it? Or across it? Blustery crosswinds bring problems – and introduce an element of luck – into archery and rifle shooting. Swirling winds in enclosed courts make it difficult to control tennis shots or, sometimes, even to throw the ball up straight for serving. Strong winds can make some golf courses, particularly seaside ones, virtually unplayable whereas without them those same links can offer little challenge. Unrelenting wind can render a football match literally a game of one half with all the play concentrated in the downwind goalmouth. It also forces lawn bowlers to adjust their delivery to allow for the wind speeding their bowl in one direction and slowing it down on the reverse end. Rough weather can cause safety problems for recreational anglers out at sea or casting difficulties for onshore fishermen. Wind of any strength can be energy sapping for those running long distances – or even walking round a golf course.

UP, UP AND AWAY
Ballooning can be dangerous when the wind blows too strongly. Competing in an endurance event from Marl, near Dusseldorf, balloonists Peter

Witman, a lawyer, and Heinrich Brachtendorf, his former instructor, were spectacularly blown over the North Sea by 75 kph winds. The gale persuaded the other competitors to put down in Rotterdam but, determined to win, the intrepid duo flew on. They were spotted by the RAF off the coast of Grimsby and intercepted by a search and rescue helicopter. The helicopter followed the balloon for several hours and a decision was taken to talk them down. They eventually landed near Carnoustie golf links at 3.30 a.m. in thick fog. Three RAF helicopters, the resources of seven police forces, radar tracking stations, coastguard teams and two lifeboats were involved in the £50,000 operation to guide them to safety. Had they not put down they would have run into a storm over the Cairngorms – an area with some of the highest wind speeds in Britain – and been unable to land until they reached the Faroe Islands, halfway between Scotland and Iceland!

Some sports, such as ballooning, gliding and sailing, depend on the wind. Too little means no sport; too much can be hazardous. In a nation that prides itself on its seafaring traditions yachting is the most popular wind sport. Unfortunately, even in Britain, the wind cannot be relied upon to supply what crews want and many a regatta proves a flop when it refuses to cooperate and blows too little or too much. In July 1976, for example, the races off Southsea for the Morgan Cup tested both the patience and endurance of crews during a hot windless weekend in which the majority of boats failed to finish in the time allowed. A few weeks later, in August, yachtsmen in the Three-quarter Ton Cup World Championships, a series of offshore races ending at Plymouth, faced gruelling conditions with strong winds that reached gale force. Two boats were dismasted and six others retired because of gear failure. Two days later, the fourth race in the programme was abandoned, not because of the gales but due to light and fickle wind, which meant that the seven-hour time limit expired before any of the boats reached the finish. The same happened the following day. But in the Enterprise Class World Championship at Weymouth in September, after four days with barely a breeze, a northerly gale blew up, turning the bay into a mass of disturbed white water and racing was abandoned.

The Beaufort Scale – devised by Admiral Sir Francis Beaufort in the nineteenth century – measures the strength of the wind at sea. Force three or four is exhilarating for the yachtsman; force five or six is challenging; force seven and above can be life-threatening.

FORCE	DESCRIPTION	WIND SPEED KM/HOUR	STATE OF THE SEA	SAILING CRITERIA
0	Calm	<1	Like a mirror.	Dinghies need to allow the sails to form aerofoil shape. Cruisers need to use engines.
1	Light air	1–5	Ripples with appearance of scales, no foam crests.	Dinghies can make forward motion with flattened sails. Cruisers still need engine power.
2	Light breeze	6–11	Wavelets, small but pronounced. Crests with glassy appearance, but do not break.	Dinghies can sail at reasonable speed with full sails. Most cruisers still need engine power.
3	Gentle breeze	12–19	Large wavelets, crests begin to break. Glassy-looking foam, occasional white horses.	Ideal conditions for dinghies with sufficient wind and small waves. Cruisers can make steady progress under sail.
4	Moderate breeze	20–28	Small waves becoming longer, frequent white horses.	Good for experienced dinghy sailors but learners should go ashore. Most cruisers can sail well.
5	Fresh breeze	29–38	Moderate waves of pronounced long form. Many white horses, some spray.	Excellent for experienced dinghy crews but others may capsize. Ideal for cruisers.
6	Strong breeze	39–49	Some large waves, extensive white foam crests, some spray.	The limit for most dinghies. Most cruisers should head for shelter.
7	Near gale	50–61	Sea heaped up, white foam from breaking waves blowing in streaks with the wind.	Dinghies should stay on shore. Most cruisers will not cope.

8	Gale	62–74	Moderately high and long waves. Crests break into spindrift, blowing foam in well-marked streaks.	All except essential crew should be below deck.
9	Strong gale	75–88	High waves, dense foam streaks in wind, wave crests topple, tumble and roll over. Spray reduces visibility.	Danger of knockdown.
10	Storm	89–102	Very high waves with long overhanging crests. Dense blowing foam, sea surface appears white. Heavy tumbling of sea, shock-like. Poor visibility.	Wave heights of 6 to 11 metres that can capsize large cruisers.
11	Violent Storm	103–117	Exceptionally high waves, sometimes concealing small and medium sized ships. Sea completely covered with long white patches of foam. Edges of wave crests blown into froth. Poor visibility.	Extreme danger.
12	Hurricane	>117	Air filled with foam and spray, sea white with driving spray. Visibility bad.	Survival is the only priority.

When winds are light, every little breeze has to be anticipated and utilised. Knowing how to find and use the wind is a major yachting skill, but its application varies according to where the sport is taking place. Ocean-going yacht racing is a considerably different test of the captain and crew from sailing on inland water or around a coastal bay. Out in the Atlantic or Pacific, skippers and navigators use their computers – not a device available to Admiral Beaufort – to download weather forecasts and barometric pressures, both indicators of likely wind shifts. These then have to be sought without taking excessive detours and losing ground to competitors.

When winds are strong many recreational sailors head for port, safety taking precedence over risking a possible death on the ocean wave. But to some the challenge of sailing in rough weather is one of the attractions of yachting, especially ocean racing. Wind that would terrify a Sunday sailor is seen as a challenge to the true seaman. Yet sometimes the weekend yachtsman has to contend with more than he bargained for, as strong wind can arrive unexpectedly, turning what started as a fun day out into a potentially dangerous situation. This occurred in the aptly named Brass Monkeys dinghy race off a freezing Leigh-on-Sea in January 1999. Hit by a sudden deterioration in the weather, all of the participating boats capsized or began to sink. There were too many in trouble for the safety vessels to cope with and all three Southend-on-Sea inshore lifeboats had to be launched. Between them, the safety crews and the lifeboatmen were able to rescue all the sailors and recover the abandoned dinghies. Even more serious was the 2001 Skandia Cowes Week Regatta, in which a combination of spring tides and blustery winds – at times reaching force 7 – led to a multitude of capsizes, sinkings, dismastings and groundings. The coastguard helicopter was scrambled 10 times, the coastguard rescue teams were asked to assist on 30 occasions, and there were 70 lifeboat launches. Again (fortunately) there was no loss of life.

Gales can occur at any time of year. Those ignorant of this fact and those who choose to ignore it can sail into trouble. August gales, for example, occur around the coasts of England and Wales on average every second or third year. Each time, as the records of the RNLI show, the winds catch out holiday boating enthusiasts, many of whom have assumed that gales do not blow in summer and have not bothered to check the coastal waters' weather forecast. In August 1984 one man drowned when a boat taking 14 trippers for a day's fishing lost its propeller in a force 8 gale off Bridlington. Of the rest, those who did not swim ashore were winched to safety by a rescue helicopter. A crewman died the same day when he fell overboard from a sailing vessel in the Needles Channel. Even seasoned yachtsmen got into difficulties that August. The Fastnet race once again ran into storms though, unlike in the tragic event five years earlier, this time no one died. (For details of the 1979 race, see Chapter 7.) Nevertheless, of the 234 starters 142 retired after a force 8 gale hit the fleet, many of them seeking shelter along the coasts of Devon and Cornwall. One yacht, the 77-foot *Drum of England*, owned by singer Simon Le Bon, capsized after its keel snapped off. The crew of 24 were rescued by a Royal Navy helicopter, six of them by a diver after they had been trapped in an air pocket in the overturned hull. The hero of the hour, Petty Officer Aircrewman Larry Slater, had previously saved eight yachtsmen from their sinking yacht in the 1979 race.

Yet strong, potentially dangerous winds can also help sailors break records. The winner of that tragic Fastnet event of 1979, *Condor of Bermuda*, eclipsed the previous best time by almost eight hours. In 1985, despite the mainsail twice being ripped in gale-force winds, Marvin Green's 83-foot *Nirvana* knocked a massive 10 hours, 20 minutes and 55 seconds off the Fastnet race record time.

Although less hazardous than ocean sailing, rowing on inland waters can also be affected by the wind, usually adversely. Britain's national watersport facility at Holme Pierrepont in Nottinghamshire is notorious for crosswinds that favour some lanes more than others. At times, rowers have to wait for a window of decent weather in order to compete on equal terms. At least conditions there have seldom been as bad as on the Thames in March 1963 when strong winds made rowing impossible. Some two dozen crews practising for the Head of the River race became waterlogged and sank. The Thames is also the site of the most well-known boat race in Britain – *the* Boat Race, the annual competition between crews from the universities of Oxford and Cambridge. Unlike rowing events in the Olympics or World Championships, this race does not take place over a straight 2 kilometres on a purpose-built laned course but on the Thames between Putney and Mortlake over a twisting, tidal course of 4 miles 374 yards. (The traditional measurement has not yet been metricated as befits such a long-established English sporting contest.) The boats are allowed to cross in front of each other and tactical steering is of the essence. On one stretch, between Craven Cottage and Hammersmith Bridge, wind and tide often clash head on. Inevitably, in the race's 173-year history the wind and other weather conditions have sometimes played a role in determining the result.

Boats have sunk on six occasions, including in 1951 when the Oxford President, having won the toss, made a disastrous mistake in choosing to start off the Surrey bank. This exposed the crew to the full force of a strong westerly wind while their opponents obtained the shelter of the Middlesex bank. Water broke over the bows of the Oxford boat in the very first strokes and it only stayed afloat for two minutes. In 1988, when Oxford again won the toss in stormy conditions, they chose the Middlesex side and this time won the race. Ten years earlier, the Cambridge boat had sunk. On the day of the race there was driving rain and howling winds. Oxford had already sunk in training and used splashboards on the boat to deflect water. Cambridge didn't. About 15 minutes into the race, Cambridge saw an opportunity to overtake but only through increasingly choppy water. As they attacked, water entered the boat at an uncontrollable rate. Oxford were also in trouble, but shipped water more slowly and finally cruised to an unopposed win, more interested in keeping afloat than recording a fast time as they were within a couple of centimetres of going under themselves.

Wind can be a propulsive force but sometimes a destructive one. On the second day of the 1973 John Player Classic at Turnberry, gusts of over 100 kph blew away most of the tented village, leaving the golfing merchandise and its drenched purveyors to cope with horizontal rain. Another tent that gave way was at the Eastbourne tennis tournament in June 1990. Julie Richardson of New Zealand received a cut on the face when a pole in the players' tent was dislodged by a fierce gust of wind. Tents might not be expected to cope with severe gales, but more permanent constructions can also be damaged when the wind gets up. In January 1976 the Bullpens fences, a feature of the Burghley Horse Trials cross-country course since it began in 1960, were destroyed by a gale. In the week before the 1987 Grand National, high winds smashed a section of the roof of the unsaddling enclosure at Aintree, snapped off a couple of the six-metre maypoles that marked the entrance to the course, and ripped spruce dressing from some of the fences.

In both these instances, there was time to repair the damage before the events took place. That option was not open to Hull City on 21 February 1996 when their home game with Chesterfield had to be called off in the interests of public safety after gale-force winds damaged the roof of the main stand at Boothferry Park. A ten-metre square piece blew off, showering the seating area reserved for visiting supporters with debris. A similar fate befell Celtic on 29 January 2000 when a Glasgow gale the previous night damaged the roof of the new Lisbon Lions stand at their Parkhead stadium and left panels flapping dangerously in the wind. Health and Safety inspectors forced the club to call off their Scottish Cup tie against Inverness Caledonian Thistle. Although the Highland club accepted that the safety of supporters was paramount, the fans were annoyed that the decision was not made before 5,000 of them had set off on the long journey south. There was also a furore because Celtic would not refund ticket money to those Inverness supporters who could not travel to the re-arranged midweek fixture, especially when it was revealed that such largesse was available to those who had purchased hospitality packages. Eventually a compromise was brokered in which Celtic offered to subsidise the coach travel for Inverness fans who came down for the second match and Inverness offered free admission to a home game for those with an unused ticket from the Celtic fixture. The Inverness supporters who ventured to Parkhead on Tuesday, 8 February, saw history made when their team beat Celtic 3–1, leading to the resignation of Celtic manager John Barnes and the wonderfully punny headline in the Scottish press of 'SUPERCALEY GO BALLISTIC. CELTIC ARE ATROCIOUS'.

Potential rather than actual wind damage can also cause problems. In November 1986 Workington FC were forced to close their ground for safety reasons. County council officials decided that the floodlight pylons, erected at Borough Park in 1915, might not be safe in strong winds. Fear of wind damage was also given as a reason why Wimbledon's Centre Court could not have a lightweight sliding roof.

Occasionally a strong wind can provide unexpected assistance to sportsmen. In January 1984 Watford's goalkeeper, Steve Sherwood, booted the ball in a high wind and saw it wafted downfield to bounce over the opposing keeper, Raddy Avramovic of Coventry, and into the net. The wind-assisted goal enabled Watford to win 2–1. Another keeper added to the score-sheet of the Leeds youth team in April 2001 when his throw was carried the length of the pitch by the wind into their opponents' goal. The most famous incident – no doubt because it was televised – came in the 1967 FA Charity Shield match on 12 August at Old Trafford, in which League champions Manchester United played FA Cup winners Tottenham Hotspur. Spurs' keeper Pat Jennings sent a mighty punt up field well into the United half. The ball sailed over all the players and was caught in the swirling high wind. It bounced over the head of a fast-retreating Alex Stepney in the Manchester goal and into the net, helping Spurs to a 3–3 draw.

Paradoxically, in some track and field events so-called wind assistance can actually work against the interests of athletes. They sometimes break records only to discover that their efforts are not recognised because the wind has been judged to have helped them too much. Since 1936, records set with tailwind assistance of over two metres a second in sprint and hurdle races up to and including 200 metres, as well as the long and triple jumps, are not ratified. Thus, despite having beaten the mark three times already, it was not till July 1995 that Jonathan Edwards officially broke the world triple jump record. On each previous occasion there had been a following wind of more than the permitted strength. Strangely, although many climatic elements can influence athletic performance, it is only one of them – wind – that is considered worth measuring in relation to potential record setting. And even here the parameters apply only to a restricted number of events. For those races around the full track it is assumed that the effects of tail and head winds will cancel each other out. Nor do the keepers of the record books take any account of how wind can affect field events such as the pole vault, discus and javelin, in which participants have to adjust their tactics or technique when it is blowing hard. This is especially true of the javelin, where a tailwind requires an increase in the angle of release proportional to the velocity of the wind and a similar decrease in the case of a headwind.

Wind can be beneficial to skiers in Scotland when the first heavy falls are accompanied – as they often are – by storm-force winds which spread the snow into every gully and depression as well as filling the gaps between the strategically placed snow fences, creating ideal skiing conditions. But when the snow cover is poor, mountain winds can blow the snow from the slopes, as it did at Glencoe in December 1995, ruining what could have been a beautiful White Christmas for skiers. Actually getting to the ski resorts and onto the slopes is not always made easy by the wind. If it blows in the wrong direction it can hamper snow-clearing operations on the access roads by depositing snow back as quickly as it is shifted. Even when the winter sports enthusiasts do arrive, safety precautions might dictate that the chairlifts and ski tows cannot operate if the winds are high.

There is a Scottish adage along the lines of 'if there isna wind it isna golf.' Perhaps Ian Woosnam might agree. In July 1996 he won the Scottish Open at Carnoustie, a tournament in which for four days the flagsticks were bent nearly horizontal. Holding one's balance in the teeth of a violent wind was a vital factor in keeping bogies off the card. Perhaps the diminutive but stocky Welshman was aided by his low centre of gravity. Lee Westwood, the young English professional, was less accommodating to the Scots and their climate and was quoted as saying, 'If anyone else tells me "this is just a wee breeze", I'll throttle them.'

Wind has always been of concern to the golfer. Modern research has shown that the flight of a golf ball can be reduced or increased by around 85 cm for every kph of wind speed. Old golfers may not have been aware of the exactitude of scientific experimentation but would have recognised the effect of wind on their game. Playing into the wind requires the ball to be kept low so that it is not held up in the air and blown off course. Playing with the wind means learning to impart spin on the ball to prevent it running on through the fairways and greens. Any mishits or errors of judgement are grossly magnified by the wind. What might have just trickled off the fairway can disappear into the jungle that some sadistic greenkeepers call rough.

Normally, golfers can cope with the speed and direction of the wind and adjust their yardage charts accordingly, but sometimes the wind is gale force and such adjustments are blown out of the window and even off the course. When it is difficult just to tee the ball, controlling its flight becomes as much a matter of hope as of technique and golf can become a farce. The influence of the wind can clearly be seen in a comparison of the winning scores of two pre-war *News Chronicle* tournaments at East Brighton. In 1937 Ernest Whitcome triumphed with 268 for the four rounds while a year later, playing in a howling gale on the same course, his brother Reg won with an aggregate of 300. The variation in scores was not attributable to differences in skill – the brothers were both Ryder Cup players – but to the influence of the strong wind.

Yet is there such a phenomenon as a wind-free round on a British golf course, particularly those near the sea? Virtually every golf instruction manual from the 1890s onwards devotes a chapter to playing in anything from a light breeze to half a gale. Harry Vardon, winner of six Open Championships, maintained that a good golfer had to have knowledge of wind conditions to rival that of a sailing vessel skipper.

The dunes that bring the distinctive character to many seaside links golf courses are created by blown sand. Unhappily, for many clubs, this is an ongoing process and they have to wage a constant battle against the wind's determination to change the physical features of their course, a costly and often frustrating experience. On one occasion, in April 1957, a sandstorm actually halted the Northern Open Championship which was being held on a links course at Nairn. A violent wind lifted sand not just from bunkers but also from the local beach. This then mixed with fertiliser blown from nearby farmland – and dust from almost everywhere – to drastically reduce visibility. The Nairn golfers could consider themselves lucky that only their tournament suffered. In October 1694, the estate and village of Culbin between Nairn and Forres vanished in a violent north-westerly sandstorm when several square kilometres of land were covered to a depth of up to 16 metres.

Then, of course, there was that 'hurricane'. The one missed by the weather forecasters. On the night of Thursday, 15 October 1987, and into the following day, one of the strongest winds ever to hit inland Britain swept through southern England, disrupting electricity supplies, ripping out 15 million trees, damaging 3 million houses and killing 18 people. Pedantically, the existence of a hurricane was denied after the event by Dr John Houghton, director general of the Meteorological Office, who insisted that there were only intense gusts of short duration rather than the high wind lasting hours that typifies a hurricane.

Inevitably sport was affected but it escaped relatively unscathed as the 'hurricane' struck midweek and had much less impact than if it had arrived at a weekend when a fuller sporting programme would have been scheduled. The first Westminster Serpentine Rowing Regatta was postponed for a day while replacement boats were found for those crushed when two trees fell across their trailer. Friday's racing at Newmarket was called off, the first weather-related abandonment since snow caused the loss of the Craven meeting in April 1908.

Workmen toiled throughout Friday – day and night – to repair the grandstand, re-erect the long stretches of running rail that had been blown over and replace the entertainment marquees that had blown down. Golf was worst hit. On the Thursday the torrential rain that preceded the wind led to the suspension of the first round of the World Matchplay Championship at Wentworth. It was completed on the Friday, although all approach roads to the course were blocked by fallen trees. Some £30,000 (£50,000 today) was lost at the gate. Elsewhere, the first Women's Professional World Team Challenge golf tournament at Broome Park near Canterbury was abandoned because of storm damage. The head greenkeeper said it would take three months to get the course back in shape. All over the South-east, courses reported damage. At Royal St George's in Kent, roofing from one clubhouse blew 45 metres and embedded itself in the 18th green, sticking up like a shark's fin! Yet, by Saturday, normal sporting service had been resumed. No league football matches were called off, all scheduled racing took place, and the World Matchplay quarter-finals went ahead only a day late.

THE BEAUFORT SCALE (ADAPTED FOR USE ON LAND)

FORCE	DESCRIPTION	WIND SPEED KPH	SPECIFICATION
0	Calm	<1	Smoke rises vertically.
1	Light air	1–5	Direction of wind shown by smoke drift but not by wind vanes.
2	Light breeze	6–11	Wind felt on face, leaves rustle, ordinary wind vane moved by wind.
3	Gentle breeze	12–19	Leaves and small twigs in constant motion, wind extends light flag.
4	Moderate breeze	20–29	Wind raises dust and loose paper, small branches move.
5	Fresh breeze	30–39	Small trees in leaf start to sway, crested wavelets on inland waters.
6	Strong breeze	40–50	Large branches in motion, whistling in telegraph wires, umbrellas used with difficulty.
7	Near gale	51–61	Whole trees in motion, inconvenient to walk against wind.
8	Gale	62–74	Twigs break from trees, difficult to walk.
9	Strong gale	75–87	Slight structural damage occurs, chimney pots and slates removed.
10	Storm	88–101	Trees uprooted, considerable structural damage occurs.
11	Violent storm	102–177	Widespread damage.
12	Hurricane	>177	Widespread damage.

3

Water, Water, Everywhere

Britain does not have a particularly wet climate, in spite of all the jokes to the contrary, but its reputation has not been helped by a recent succession of wet winters . . . and springs . . . and autumns. In the last 50 years, the average annual rainfall in England and Wales has been around 900 mm, with significantly higher amounts in the west and north and drier conditions in the south and east. London has lower average rainfall than Rome or Nice; so, unbelievably, does Manchester. Scotland also divides in a similar fashion: Glasgow is much wetter than St Andrews; hilly areas such as the Grampians are twice as wet as Glasgow. Fortunately, most sport takes place at low levels. Unfortunately, many important annual events are scheduled for summer in the southern half of England, the time and place of the heaviest short downpours and thunderstorms, which helps to explain why Ascot, Wimbledon, Lord's, Silverstone and Epsom can be rapidly washed out. This chapter may cause you to smile (or weep) in disbelief at the amount of disruption that wet weather has caused sport.

Two of the above venues and the sports they represent, cricket at Lord's and tennis at Wimbledon, are highly susceptible to rain. Equipment and playing surfaces quickly become too wet for matches to continue in safety and players usually come off if light rain or drizzle persists for longer than a few minutes. Racing cars are equipped with wet weather tyres for better grip in wet conditions but standing water on the track and heavy rain or spray which obscures visibility can result in a race being stopped. Sudden downpours at racecourses may rapidly alter the conditions of the turf and drench horses and jockeys but they seldom lead to abandonment. Thunderstorms and the risk of lightning are a more serious matter.

RAINDROPS KEEP FALLING . . .

A rain day is a 24-hour period in which 0.2 mm falls; a wet day is one with 1 mm or more. The size of the raindrops determines whether the wet stuff is classified as drizzle or rain. The intensity of rainfall is calculated according to the amount per hour – more than 4 mm per hour is heavy rain, less than 0.5 mm per hour is light, with moderate falls in between.

Another form of precipitation, sometimes associated with thunderstorms, is hail. Although they are ice droplets, the largest hailstones capable of damaging crops, cars and greenhouses tend to fall in the warmer half of the year, usually in the southern counties of England. When a combination of torrential rain and hail hit Epsom during the Derby meeting of 1979, the Coronation Cup had to be postponed.

Summer Showers

Wimbledon has a weather folklore all of its own and merits a separate chapter. Old Trafford is another venue forever associated with a four-letter word – rain – and will be examined later. It is not alone, however, in contributing to that great English institution, the wet cricket match. Statistics in *Wisden* suggest that there are numerous candidates for the prize of worst cricket season of the twentieth century. Calculations using several measurements (such as the amount of rainfall from May to August, or the number of batsmen averaging over 50 and bowlers under 20 runs per wicket) highlight 3 bad years in the '50s – 1954, 1956 and 1958 – and a succession of wet summers in the period 1930–32. But the top honour goes to 1903, with 1924 a close second. The winning year saw a rainfall figure of 450 mm in four months, during which 44 bowlers averaged fewer than 20 runs and only 2 batsmen achieved an average of over 50.

It would be impossible to do justice to the subject of rain and cricket in a few brief paragraphs. There are so many possible scenarios: the complete wash-out, in which no ball is bowled; the rain-affected one-day match with all the ramifications of reduced overs; the tactics required to keep players on the field when combinations of rain, low cloud and bad light conspire to end play early; the extraordinary variety of wickets that teams were asked to play on in the days before pitch covering; the championships won and lost as a result of rain delays. Some might argue that the topic justifies a book of its own, but they will just have to make do with a few examples here.

In June 1907, Northants bowled out their visitors, Gloucestershire, for 60 on the second day of a county match. Celebrations, however, were quickly forgotten. The pitch was soggy and the home side soon found themselves in dire straits, dismissed for 12 runs, the lowest total in first-class cricket during the twentieth century. They were spared utter humiliation when rain washed out play on the final day and allowed them a draw. The Midlands witnessed another rainy episode in August 1967 in the second Test between England and Pakistan at Trent Bridge. England had just started their innings on the first day, having bowled out Pakistan for 140, when a violent thunderstorm struck the area around 5 p.m. The field rapidly became a lake and the fire brigade was called on to pump over 450,000 litres of water from the ground. Amazingly, play started at 12.45 p.m. on the second day and, even with no play possible on day four, England ran out comfortable winners by 10 wickets.

Pakistan were the unfortunate visitors again in 1978 when the third Test at Headingley resulted in a draw. Rain wiped out 12½ hours play, and an entire day was abandoned because the square and bowlers' foot marks had become saturated, the result of inadequate pitch covering. While nearby matches went ahead unaffected by weather, the Test teams kicked their heels all day waiting for the wicket to dry out. Pakistan's first innings, begun at 5 p.m. on the first day, finally ended at tea on the fourth. The match, needless to say, was drawn. The final word, as ever, belongs to umpire Dickie Bird. He recalls a one-day international between England and Australia at The Oval in 1977 when play was halted for an hour by torrential rain. Although conditions were atrocious, the two captains, Mike Brearley and Greg Chappell, agreed to play to a finish because of the Silver Jubilee celebrations taking place in London the following day. Dickie described conditions as the worst he had ever experienced, with rain, hail and even sleet leading to pools of water on the outfield and players slipping and sliding on the wet grass. Australia won.

Rain on the racecourse is miserable for everyone, particularly at the traditional summer 'picnic' meetings for the Derby and Royal Ascot when vast crowds gather on Epsom Downs and Ascot Heath. There is no escaping the downpours on these exposed tracts of land; the same can be said for other sports which take place beyond the confines of a stadium or some form of shelter. Equestrian events, golf tournaments, motor sports and other competitions which are held in 'the great outdoors' can easily become trials of endurance for contestants and spectators alike. The British Grand Prix at Silverstone has often experienced torrential rain. It was wet in 1988 when Nigel Mansell took second place behind Ayrton Senna and wetter still in 1998 when both David Coulthard and Damon Hill spun off the slippery track. In 1993 the qualifying sessions took place on a circuit several centimetres deep in water, although the race itself was dry. Those who criticised the notorious April staging of the event in 2000 might care to remember the poor weather suffered by the race in its normal summer slot.

ASCOT OPENING DAY?

Although racing takes place there all year round, the Berkshire course is synonymous with Royal Ascot, the highpoint of the flat racing season for quality horses, Queens . . . and queer hats. Poor Ascot has had its share of dreadful weather at this June meeting. Storms and lightning twice led to fatalities (see Chapter 7) but torrential rain has upset proceedings on numerous occasions. The Royal carriage procession up the course was shortened in 1954 because of wet ground, cancelled in 1964 and took place in 1977 with the wheels of the open landaus visibly sinking into the mire. On that occasion nearly 40 mm of rain had fallen in 36 hours and the going was officially described as 'soft'. The meeting was hit by a rainstorm and high winds in 1992, after a glorious start to the week, and the first day in 2000 was a meteorological disaster; supplies of umbrellas were apparently exhausted by 10.30 a.m. and planned outfits had to be hastily re-arranged. The worst wash-out, however, occurred in 1964 when two whole days were lost to the British weather. It had rained a little on the opening day, but Wednesday saw the start of a lengthy downpour which hardly stopped for the next 24 hours. Thursday's racecard was eventually called off at 3.15 p.m., by which time an American visitor described a scene in which the elegant racegoers sat 'dripping steadily, like trees in a rain forest'. The course never dried out after this onslaught and racing was abandoned for the remainder of the week. The episode led to an extensive drainage programme at the course in which 45 kilometres of pipes were laid. It also put paid to the record-breaking exploits of the nine-year-old Trelawny. Winner in 1962 and 1963 of both the Ascot Stakes over 2½ miles and the Queen Alexandra Stakes, the longest race in the flat racing calendar at 2 miles 6 furlongs, he was attempting to do the double again in 1964. He finished second in the Ascot Stakes, however, and the Queen Alexandra was never run, which was unfortunate as Trelawny was the only entrant and would have obtained a walkover.

The Great British Bank Holiday

Sometimes it seems that the very mention of the words 'bank holiday' are enough to conjure up dark clouds on the horizon. This can be disastrous for sport, a great deal of which is crammed into those nationwide breaks. Easters tend to be unpredictable because they move within a four-week slot in the spring calendar, and the British climate is so variable that a late April holiday holds no more guarantee of warm sun than four days in the previous month. Easter 2002 fell on the final weekend of March and produced beautiful, sunny weather. Easter 2000 was celebrated three weeks later and formed part of the wettest April since 1756. The start of the cricket season was ruined, canoeists capsized on swollen rivers, horse trials throughout the country were postponed and race meetings abandoned. The 25-strong fleet in the Tall Ships race from the south of England to Spain was hampered by bad weather and rough seas; six vessels dropped out because they were taking on water and their crews were seasick. The weekend is best remembered for the infamous British Grand Prix at Silverstone where the circuit was partially flooded on practice day, the car parks turned into hectares of mud and would-be spectators were caught up in traffic jams that snaked across half of Northamptonshire. (For more detail on the plight of Formula One fans, see Chapter 17.) Although conditions for the race were acceptable, there were still large quantities of surface water. Rubens Barrichello spun twice and splashed through two enormous puddles on his way to the pits before retiring with 'hydraulic problems', a polite way of saying that his car was full of water!

May bank holidays fare little better. Weather expert Philip Eden has calculated that the first 17 May Day holidays (1978–94) resulted in only one full weekend of dry, sunny weather. Many were dull and cold, some were passable, others were simply wet. The late spring holiday which takes place on the last weekend of May should be slightly more reliable, according to the calendar, but often suffers similar dismal weather. In 1981 it sank without trace. Racing at Sandown and Fontwell, and polo at Windsor, were all abandoned due to waterlogging. The Windsor Horse Trials, already postponed and rescheduled for June, were finally cancelled 'due to continuous rain in Berkshire'. Prolonged rain before the Devon County show at Exeter led to the transformation of the showground into an arena of mud after heavy vehicles had churned up the surface: the show jumping was abandoned. The year 1984 also witnessed incessant rain and temperatures below 10°C in south-east England and the Midlands. There was no play in three cricket County Championship matches and the Windsor Horse Trials took place in mud-bath conditions, but at least they were completed. In late May 2000 only the dressage phase went ahead, the remainder of the competition falling foul of the relentless rain which left pools of water across Windsor Great Park. One of the wettest May weekends in recent history also saw widespread flooding in southern England, and the Volvo PGA tournament at Wentworth carried over to Monday. The winner, Colin Montgomerie, played only seven holes of his first round on Friday before rain intervened. He returned at 7.30 a.m. on Saturday to complete it but the start was delayed until 11 a.m. to allow groundstaff to squeegee and hand-mow greens and drain bunkers. It was 5.15 p.m. on Sunday when he embarked on his third round, 60 mm of rain having now fallen in 48 hours; he only reached the 13th before darkness fell. Montgomerie was up at first light again on Monday to finish the round before setting out on the final 18 holes. With a winning total of 271 (67, 65, 70, 69), the weather had no detrimental impact on his golf – though he described conditions as 'unreal' – but the head greenkeeper could not say the same about the Wentworth course. He had never seen weather like it in 25 years on the job and was concerned that some nearby fairways, doubling up as car parks, were beyond immediate repair.

Finally, of course, there is the late summer holiday at the end of August, the final public holiday before Christmas and the culmination of the great British summer. Not in 1986 it wasn't! Rain and gale-force winds on Monday washed out four county cricket matches without a ball being bowled and severely disrupted the remainder. The World Formula Two Powerboating Grand Prix at the Royal Victoria Dock, London, was halted because driving rain made conditions too hazardous. Crowds at the Notting Hill Carnival were reduced to less than a third of those expected. In Birmingham, Britain's first attempt to stage motor racing on public roads had to be abandoned during torrential rain. A four-kilometre circuit near the city centre had been prepared for a two-day event, the highlight of which was to be a Formula 3000 race, but the heavy rain and spray sharply reduced visibility. The race was called off on safety grounds after 25 laps.

Wintry Water
In spite of the terrible weather on that weekend, rugby and football programmes went ahead as normal; it takes a lot of rain or poor drainage to force the

abandonment of the traditional winter games. Precipitation has to result in waterlogging, standing water on the pitch or flooding before these sports will be called off and pre-fixture inspections usually ensure that this happens before the start. But occasionally the weather and ground conditions can deteriorate so rapidly that continuation is impossible. The first leg of the Inter City Fairs (now the UEFA) Cup final between Juventus and Leeds had to be abandoned after 51 minutes in May 1971 because hours of torrential rain had turned the pitch into a lake. Leeds went on take the trophy on the away goals rule, having drawn 2–2 in Turin and 1–1 at Elland Road. Conditions for horse-racing can also worsen dramatically in a short time. The authors endured an afternoon of horizontal rain and wind in October 2000 at Leicester racecourse, where the clay soil often makes the going heavy, before being mercifully released by the abandonment of the meeting. The ground, already sodden after a wet autumn, had become waterlogged in the space of a few hours. Even before that, conditions were extremely unpleasant for all: jockeys, horses, officials and the few hundred miserable spectators who had braved the day.

Postponement due to flooding was a close-run thing when the final of the Rugby League Challenge Cup was held at Murrayfield in the abysmal spring of 2000. Two days before the match the ground was under a metre of water as a result of the nearby Water of Leith bursting its banks after 48 hours of incessant rain. Groundstaff were allegedly pumping water out of the stadium at a rate of 1,800 litres an hour, but the game went ahead as scheduled on good ground thanks to human sweat and excellent drainage. At the opposite end of the sports business, non-League Gravesend had to postpone all their home ties for the foreseeable future in November 1986 because heavy rain and the proximity of the River Thames had seen their pitch deteriorate into a sea of mud 30 cm deep in places. The water table had been raised by a combination of a wet autumn and the closure of several nearby factories which had previously consumed large quantities of river water.

A few unlucky racecourses – Newton Abbot, Nottingham, Perth, Windsor and Worcester – also have the misfortune to be situated on the banks of rivers. Perth lost its flagship three-day meeting in two successive Aprils (1999 and 2000) but it went ahead in 2001 in spite of persistent rain. Windsor racecourse, alongside the Thames, used to suffer regular flooding and consequent abandonments in the days when it ran meetings throughout the year. Now it concentrates on summer flat racing but still lost an April race day to waterlogging in 2001. Worcester, submerged by the neighbouring River Severn at frequent intervals, is renowned for its soft going and constant risk of abandonments. When it lost a meeting in January 1994, the general manager ruefully recounted that the Worcester rowing eight had been seen racing down the course, screaming 'photograph!' as they passed the winning post!

Even if a ground threatened by waterlogging survives a pitch or course inspection, there is likely to be a considerable impact on the quality and nature of what follows. It is possible to race and play skilfully in pouring rain providing the downpours can be absorbed; it is less easy to produce fast, entertaining sport in heavy going. The appalling pitch at Wembley in the late 1960s – resembling 'a cabbage patch' in the words of Manchester City manager Joe Mercer – was probably a factor in the downfall of Arsenal when they faced lowly Swindon Town in the final of the League Cup in 1969. The Third Division side coped better with the Wembley mud and won 3–1 in extra time. When Mercer's men won the FA Cup two months later, the playing surface did nothing to assist their smooth passing game and should have played into the hands of their opponents, Leicester. Manchester still managed a 1–0 victory. Rugby, though often played in conditions ranging from mud to deep mud, is another game that benefits from a dry pitch. It seldom gets one, particularly towards the end of a wet winter. The Gala Sevens in April 1979, after months of snow and rain, took place in vile weather on a sticky, churned-up pitch. When the ball was kicked ahead, it failed to skid across the ground but simply stuck in the mud, like an oversized dumpling.

The muddiest and most debilitating steeplechases usually produce slow times and, often, unlikely winners. Heavy going tends to militate against lightweight or classy horses with a turn of speed and promotes the chances of honest, one-paced plodders. (The same might be said of football!) The 2001 Grand National is a case in point. The atrocious conditions – pouring rain and bottomless ground with deep divot holes filled with water – led to the slowest time for over a century, only 4 finishers from a field of 40 and a winner, Red Marauder, who was reckoned to be one of the worst jumpers to win a National, and that was the opinion of his jockey! However, the controversy over whether the race should have been run ignores the many other Grand Nationals which have gone ahead in appalling weather and ground conditions. Only one horse finished in 1911; very heavy going was reported in 1928 and 1947; it was so muddy in 1955 that the water jump, situated in the most waterlogged area in front of the stands, was omitted from the race; and some jockeys thought that the slog in the mud of 2001 was not as bad as 1998, another slow, mud-bath epic won by Earth Summit.

Sometimes the sheer class of the athlete triumphs over the elements. In equine terms, Desert Orchid showed his ability by winning the Cheltenham Gold Cup in 1989 on a left-handed course (which he disliked) and on heavy going (which he hated even more). The final day of the National Hunt Festival was very much in doubt, with early morning snow turning to steady rain, and fire engines were drafted in to pump water off the course. The winning jockey, Simon Sherwood, admitted later that Dessie had loathed the ground but his courage had brought him victory against the odds. The same could probably have been said of West Bromwich Albion in the 1931 FA Cup final. On a day of strong wind, driving rain and slippery turf, they overcame Birmingham 2–1 in a match of high quality, considering the conditions. Three years earlier, Scotland had beaten England 5–1 at Wembley on another day of pouring rain. The Scots had displayed great ball control and skill in spite of the relentless downpour. It is worth remembering that the leather football used at that time simply absorbed water; to kick it 60 metres was an amazing feat on a wet day.

Rainy conditions leading to greasy surfaces underfoot and slippery balls in handling games such as rugby certainly bring luck into the equation; sometimes they result in unexpected thrills and spills on the pitch, entertaining for spectators but unfortunate for players. The Rugby League Challenge Cup final of 1968 will be remembered for the atrocious weather in which it was played and a missed kick in the dying seconds of the game which would have given victory to Wakefield Trinity. Instead, in a match dominated by the elements and the sodden ground, Leeds ran out 11–10 winners. In normal conditions, Leeds would have been expected to win more comfortably but players easily lost their balance and often slid and slithered for several metres in the mud. It was not, could not be, a classic game of rugby.

Soggy Springs and Awful Autumns

You probably don't want to hear any more instances of rain wrecking sport. Having dealt with winters, summers and bank holidays, surely that's enough? Oh no, you don't get away without experiencing some truly dismal springs and autumns – like May 1994 when Italy won a controversial show jumping Nations Cup at Hickstead. The appallingly wet conditions meant that none of the horses and riders escaped with fewer than eight faults in the first half of the competition and Nick Skelton retired after a crashing fall, refusing to risk his horse again on the heavy going. Team managers asked for fences to be lowered and the course shortened for the second round, and the Italians finally came out on top. The same day saw the cancellation of the one-day cricket international between England and New Zealand and, according to *The Times* correspondent, the English women's amateur golf championship was watched by 'a hardy band of sodden spectators and damp dogs'.

And what about May 1963, when the very first match of the new one-day Gillette Cup series between Lancashire and Leicestershire at Old Trafford was affected by rain and had to be extended into a second day? (The original stumbling block to a knock-out competition had actually been the issue of expense if matches were interrupted and forced into this situation. It was only when Gillette agreed to underwrite the tournament that it got off the ground.) Then there was spring 1998 when, by 4 May, Kent county cricket players had still not managed any outdoor net practice because of rain, and Easter racing was cancelled at Ludlow, Newton Abbot and Towcester, the last because the course was flooded. Whoever decided that mid-April was a good time to stage the opening round of cricket's Benson & Hedges Cup got the chaos they deserved in 2000 when six matches were abandoned due to the weather. The semi-finals at the end of May resulted in one complete wash-out and the other reduced to 25 overs. Finally, 1959, which turned out to be a warm, dry summer, had started with a very wet spring. During the Badminton Horse Trials in April, the continuous rain during the dressage led to one horse being unable to perform a 'halt and rein-back' movement because, once it had halted, it couldn't pull its legs out of the deep mud in the arena to step backwards! The cross-country course had to be drastically altered, with modifications to four fences and five removed completely. In 1988, after three total cancellations of the event, it was decided to move it from April to May in the hope of securing better weather and drier ground.

Autumns are more settled – allegedly – but not in October 2000. The Rugby League World Cup was plagued with teeming rain, howling winds and consequently a lot of empty seats, and the Solheim Cup between European and American women golfers at Loch Lomond in the same month saw three holes virtually submerged under rainwater. At tea-time on Saturday, there was a question mark over whether the tournament could be completed. (It was – Europe won.) Scotland was not alone in its sodden misery. The World Matchplay golf at Wentworth over the same weekend was also hit by torrential rain, but with hail and lightning thrown in for good measure. In 1968, a wet autumn saw water being pumped off the Knavesmire racecourse at York for its final meeting of the year and the same golf tournament struggling to take place over a flooded West course. On the day before the event many of the greens were just ponds amongst the trees. The semi-final between Tony Jacklin and Gary Player took place in unrelenting gloom, rain and a brisk wind on a course awash with casual water, muddy lies, sopping greens and fairways heel-marked by the crowd. They were all square after 36 holes and Jacklin, who had played a superb inward nine to draw level, wanted to continue. But failing light meant a resumption next morning instead, when he lost the first play-off hole. Player went on to win the championship.

You want to hear more? We thought not.

4

The Heat is On

But not very often! Any chapter about the problems of playing sport in the heat of a British summer will necessarily be short!

The British have an odd relationship with the sun. Millions leave these shores in search of it – winter sun, Easter breaks, summer fortnights – but when it arrives on our doorstep, we seem to find it disconcerting. If the temperature rises above 25°C, it's hot. If it stays like that for several days, it's a heat wave. If it lasts for a week, many Brits begin to complain of overheated cars and trains, shops and offices; of parched lawns and wilting flowers; of crowded beaches, parks and swimming areas. Choruses of 'we're not accustomed to this' rise from pavement conversations, particularly from the middle-aged and elderly, together with disapproving comments and looks as youth collectively strips off as much as it decently can and disports itself flagrantly in public. Make the mistake of holidaying abroad during one of these British heat waves and you'll be treated, gloatingly, to a discourse on your return as to how wonderful it's been at home and what a pity you wasted your money. If your weather was less than perfect, don't admit it!

One of the difficulties with the British 'summer' is that it is just as likely to appear, briefly, in April or May as in its proper slot in the calendar. There have been heat waves during the London Marathon in April and end-of-season rugby tournaments in May. The 1976 Middlesex Sevens were watched in sweltering heat by a crowd of 60,000, half allegedly stripped to the waist (the male half, one assumes!). Loughborough beat Harlequins 21–20 in the final. On 3 April 1995, one marathon runner died and thousands suffered from heat exhaustion as the London Marathon was run in temperatures of 21°C, one of the hottest April days for many years. In a typical rollercoaster episode in British weather lore, competitors had been training in snow the previous week! Over 4,700 of the 26,000 runners required treatment, mostly for dehydration, and 40 were taken to hospital. The following year, it reached 21°C again. Liz McColgan won the women's event and Mexican Dionicio Ceron was the men's champion for the third successive year, beating Belgian Vincent Rousseau (who hated competing in warm weather) into second place.

A spell of dry, warm weather in spring can have far-reaching consequences for horse-racing and country pursuits. The 1990 Grand National took place under these conditions and resulted in a win for Mr Frisk in record time. The exceptionally fast ground, dried out by sun and above average temperatures, led to a fast and furious pace and a winning time of 8 minutes 47 seconds, the first and only Grand National to have broken 9 minutes. (The previous record, held by Red Rum but made possible by the stunning performance of the beaten Crisp in 1973, was also achieved on fast, firm ground.) Two horses were withdrawn at the last minute because their trainers thought the ground unsuitable and there were seven equine fatalities at the three-day meeting, for which the lightning-fast conditions were held largely responsible. Hard spring ground always leads to small fields for point-to-point races and warm weather has sometimes put a sudden end to the hunting season. According to a spokesman for the British Field Sports Society, hot dry spells 'do rotten things for scent'! Hunting people prefer nasty, gloomy overcast days.

Of course, there have been glorious summer days of sport in the appropriate months, although even they can sometimes bring problems for competitors and venues. The Empire Games, held in Cardiff in July 1958, were blessed with sunshine and sometimes sultry heat. When Dave Power of Australia won the 6-mile event (the equivalent of the 10,000 metres) he collapsed after crossing the line but recovered sufficiently to win the marathon on another day of sticky afternoon sun. During the Open golf championship at Turnberry in July 1977, the rough perished in the heat, there was no wind and golfers took the course apart; Tom Watson won with a record score of 268. Cycle road races on hot days often result in the use of hosepipes to cool competitors and more cases of sunburn than any other medical complaint. During the Kelloggs' Tour of Britain in August 1990, temperatures touched 37°C in Yorkshire as the riders pedalled from Hull to Sheffield. Police warned that road surfaces were melting and the unfortunate cyclists who were required to give drug control samples found that they had to drink more than two litres of water to produce the necessary result. Footballers who toiled in the heat of Wembley at the end of the season in May sometimes found themselves playing again two months later at the height of summer. Tournaments such as the Anglo–Italian Cup and matches like the Charity Shield are frequently slugged out by half-fit players in blazing sunshine.

Cricket, the quintessential English summer sport famed for its rainy interruptions, has also had its halcyon days of sunshine, stroke play and warm beer. Much cricket history focuses on legendary beautiful summers, such as 1921, 1934 and 1947, but other years have produced unusual cricketing conditions. A match between Surrey and Lancashire at The Oval in July 1888 was stopped for an hour after lunch because the umpires apparently judged it too hot to continue. Amazingly, the first 60-over cricket World Cup coincided with a sunny June fortnight in 1975. The hot summer of 1990 saw 10 batsmen score over 2,000 runs in county cricket and another match between Surrey and Lancashire resulted in the home side reaching 707 for 9. The opposition replied with 863. But hot summer weather does not always favour the locals. Allan Border's Australian team reclaimed the Ashes in 1989 on sun-baked pitches which suited the visitors. Two Tests were drawn because of rain, at Edgbaston and The Oval, but Australia won the remaining four matches in convincing style – at Headingley by 210 runs, at Lord's by 6 wickets, at Old Trafford by 9 wickets and at Trent Bridge by an innings and 180 runs. It had been 40 years since the Australians, under captain Donald Bradman, had achieved 4 Test match victories on English soil.

Glare from the sun can also be a problem for cricketers as it bounces off glass and metal surfaces. A match between Essex and Derbyshire had to be halted for 15 minutes in 1963 to allow cars to be moved as the light was reflecting from windscreens. In a well-known incident during the 1995 Test between England and the West Indies at Old Trafford, umpire Dickie Bird marched down the wicket shouting and gesticulating at the hospitality boxes, from which reflected light was blinding players on the field.

Hot dry summers have also resulted in drought conditions. The infamous summer of 1976 merits a chapter of its own but there have been other years in which water shortages have affected sport. Spring 1984 saw one of the warmest Easters for 40 years and, by July, there were hosepipe bans in operation across Wales and the West Country. Sports clubs in parts of Devon and Cornwall were worried that they would soon be unable to water pitches, as had happened in 1976. There was, therefore, considerable anger in the local community when it was discovered that a ski centre in Plymouth was using thousands of litres of water to ice its artificial slopes. It was able to take advantage of a legal loophole because indoor ski slopes were not covered by the 1976 Drought Act. The year 1989 also saw conditions reminiscent of 1976. Parched cricket grounds began to crack, racehorses were withdrawn from meetings as trainers decided not to run risks with their charges' welfare and anglers were faced with low water levels in rivers and canals.

The quality of a British summer can apparently be judged by the sales figures for Pimm's, an alcoholic concoction much favoured at those archetypal English summer venues, Wimbledon, Ascot and Henley. Wimbledon was rain free and hot in 1976 and 1993. Nearly 400 spectators were overcome by heat on the first Saturday of the championships in 1976. The sprinklers were out every night in 1993, applying 1,400 litres of water to the hallowed Centre Court turf in an effort to keep it green during what turned out to be the best fortnight in a coolish summer. Even 2001 was blessed with good weather – and 20 per cent higher sales of Pimm's than the previous year – until the last fateful Friday spoiled the tournament for fans of Tim Henman. At Royal Ascot, the weather seldom seems to do anything by halves. Heavy showers, gusty winds and thunderstorms have been matched by heat waves and hard ground. In 1992 the dry, warm weather led to a rash of record times, a reflection of the firm going as much as the quality of the horses. (Two days later the Royal carriage procession was greeted by flashes of lightning and torrential rain.) Some 11,000 bottles of Pimm's were consumed in 1998. In 2000 a scorching June weekend left the course so fast and firm ahead of the meeting that some trainers threatened to remove top horses rather than risk injury. The first day of the Henley Regatta in 2001 saw temperatures of nearly 30°C and many hot-under-the-collar spectators as officials refused to allow them to remove their jackets in the stewards' enclosure. As a concession, they were permitted to sit in shirtsleeves in the lunch tent, where no doubt Pimm's and the obligatory champagne featured prominently on the menu.

Cheers! Let's drink to a glorious British summer!

5

Frosty the Snowman

ew sports can cope with snow. To be more accurate, the quality of sporting performances is usually reduced by snow-covered surfaces and whirling snowflakes. Yet football matches have often gone ahead on pure white pitches, with Day-Glo orange balls as the sole concession to players and public. Scantily clad footballers in Britain have risked fractured limbs and frostbitten extremities in the name of entertainment while their more sensible north European colleagues were, like the UEFA tournaments, taking a winter break. Snow, however, is not a guaranteed element in the British weather cycle and the post-war years have seen very few harsh winters. One of the worst occurred in 1947, a remarkable year in terms of the amount that fell and the very cold temperatures that kept it lying around from mid-January to mid-March. By then, over 200 football matches had been postponed. With a government ban on midweek games in the interests of post-war productivity and a national fuel crisis, there were fears that the league championships might not be completed. The season was extended to 14 June to clear the backlog and Liverpool, despite playing their last four games away, still managed to top the First Division. Racing was hard hit, with more days lost than in any previous year of the century, and the Waterloo Cup, the premier hare coursing event, was abandoned for the first time in its 110-year history, after three postponements.

The years 1947 and 1963 (which was so dire that it merits a chapter all to itself) were exceptional winters and the past 30 years have seen nothing comparable. Instead, there have been numerous cold snaps within averagely chilly seasons. In 1987, a particularly disruptive fortnight in mid-January not only brought sporting action to a standstill but caused the wheels of sport's governing bodies to seize up. In a vicious five-day spell of heavy snow and frost, a Scottish Football Association council meeting had to be cancelled, the first in 30 years, and both the Scottish Amateur Rowing Association and the Rugby Football Union (RFU) were unable to meet because of the weather. The opening Five Nations rugby fixtures were postponed, prompting questions about whether they should be played so early in the year. (Can the weather be guaranteed at *any* time during the rugby season?) There were suggestions that football was better suited to summer than to the British winter as only ten matches in England and Scotland were played on 17 January. The top clubs headed for sunny Spain to continue training – apart from Watford. They flew to . . . Iceland, where they played a Reykjavik select, indoors on a synthetic pitch.

The main problem with similar periods of bad weather is that they are unpredictable, unevenly distributed throughout the country and generally short-lived. They are seldom serious enough to warrant a lengthy sporting shut-down, from either an administrative or a commercial point of view, but are invariably greeted by calls for a seasonal break. The Scottish Premier League (SPL) bit the bullet in January 2000, introducing a three-week suspension, the first in British football, but the likelihood of bad weather was only one of a number of reasons for the decision. Giving players and pitches a rest were seen as equally important. The chief executive of the SPL acknowledged that there could be wintry spells in any month from November to March and admitted that if February happened to be worse than January, 'it won't look so good'. (The SPL was spared any embarrassment that year as February weather was relatively trouble free. The following winter, however, when the January shut-down was repeated, a period of snow early the following month put paid to several SPL fixtures.)

The issue of a midwinter break was also discussed in England, as it is whenever bad weather strikes. As things stand, it would be very difficult to implement during a league season in which teams average over 40 matches, together with numerous knockout competitions. In those parts of Europe where football goes into winter hibernation, there are smaller leagues and fewer games. Therefore, unless English football embarks on a radical restructuring and downsizing, fixture congestion caused by bad weather postponements is likely to remain. In Scotland, the chance of a winter break was welcomed by Premier League players, not surprisingly as their clubs paid for jaunts to Spain, Portugal or Florida. There was much talk of recharging batteries and opportunities to recover from injury. Meanwhile, the lower leagues struggled on through the usual mixture of frost, snow and postponed matches. Premier League fans were faced with stark choices – a trip to the likes of Dumbarton or Brechin to get their Saturday football fix, or an afternoon's shopping or DIY.

WINTER WONDERLAND

Snow is formed from ice crystals which, when they collide and freeze together, become snowflakes. At temperatures around 0°C, they fall as 'wet snow', the typical snowball material of most British winters. When it is colder, the crystals fail to bond and remain small, fine and powdery, the 'wrong sort of snow', popularised by British Rail in 1991. On low ground and coastal areas throughout Britain, the average winter sees only 5–10 days of lying snow, but this increases with height above and distance from the sea. There may be 20 snow days in parts of central or northern England, and around 70 at the Drumochter Pass (460 metres above sea level) on the A9 north to Inverness. At the top of Ben Nevis, snow might lie for over six months.

Blizzards occur when snow and strong winds coincide. The wind can pile up snow against any obstruction, causing deep drifts and obliterating landmarks. In hilly areas, blizzards can result in treacherous conditions for walkers, not only during heavy snowfalls but also when winds whip already-fallen snow particles back into the air, leading to very poor visibility and disorientation. The onset of blizzard conditions can be very rapid, catching out skiers and snowboarders as well as climbers and walkers.

Frost, the other wrecker of winter sport, occurs when air and ground temperatures fall to below 0ºC and a layer of white, feathery ice crystals forms on exposed surfaces. The greatest damage to sports venues, such as football pitches and racecourses, takes place when the drop in temperature causes soil moisture to freeze to a depth of several centimetres. Usually this happens overnight, and if the increased warmth of the day fails to melt the frost from the ground, or an inspection shows that the frost has penetrated too deeply for a rapid thaw to occur, fixtures will be called off.

Almost every spell of snow that lasts for more than a few days is labelled 'the big freeze' or 'the big chill' by a press which likes to sensationalise any weather event, while the ubiquitous headline 'the weather won' is a favourite in the sports sections. Changeable winter weather and the frequency of mild winters in Britain seem to have encouraged such complacency amongst press and public that when snow does fall, disbelief, outrage and hyped-up headlines always seem to accompany the blizzards. Anger is invariably directed towards weather forecasters, council officials and railway operators, as if they had no right to let it happen! Perhaps this is not surprising when the British transport network descends so rapidly into chaos and the level of disruption seems totally disproportionate to the actual amount of snow cover. Hard, snowy winters or winters with just a few snowy days have one thing in common: they play havoc with the fixture lists. The third and fourth rounds of the FA Cup, traditionally held in January, often take a battering. The winter of 1979 saw a month go by before the third round was completed. In 1993, 10 of the 32 ties in the FA Cup third round were postponed. In 1997, 15 were called off, although neither of these seasons can compare with January 1963, when only 3 matches went ahead on the appointed day. In 1985, the Cup escaped major disruption, but the league programme the following week was hit hard, with 35 games off in England and Scotland. The year 1996 saw snow and ice wipe out all but three of the FA Cup fourth-round ties.

If football matches go ahead in bad weather, opinions as to whether they should have been played usually depend on the outcome. With luck playing its part, skill reduced to a minimum and players performing like novice ice skaters, the situation is often ripe for giant-killing. When referee Mike Reed judged that the third-round match at Second Division Wrexham should go ahead in January 1997, as the pitch was unlikely to endanger limbs, Harry Redknapp, the manager of visiting Premiership side West Ham described the ruling as 'scandalous'. The game, played on a snowbound surface with blue touch-lines, ended in a 1–1 draw with Redknapp repeating his criticism of the decision to play and suggesting that the snow had prevented the true run of the ball. Conditions were certainly not ideal, particularly for the visitors. One overseas member of the West Ham team had never even seen snow before, let alone played on it! Worse was to follow. The replay at Upton Park was called off less than two hours before kick-off because of fog and, when it finally went ahead, West Ham were beaten by a single goal.

On the same third-round Saturday, referee Graham Poll ruled that the match between Reading of Division One, and Premiership side Southampton was also playable. Reading goal scorer Darren Caskey agreed with the decision, saying that there was nothing wrong with the pitch – after his team had won 3–1. Not surprisingly, Southampton manager Graeme Souness saw it differently. He thought the frozen surface was dangerous but, on voicing this opinion to the referee before the match, he was allegedly told that it would be just like a hard pitch in August and if the teams played at 90 per cent, they would be OK. (The problem with that line of thinking, of course, is that cup ties are usually played at 110 per cent!) Players donned trainers in an effort to maintain balance on the frosty surface while fans were confronted with chemical toilets as all the pipes at the ground were frozen. Another Premiership outfit bit the dust. (Or should that read 'the frost'?)

A classic weather-related FA Cup drama took place in the third round in 1972. Newcastle of the First Division were set to meet Southern League Hereford at St James' Park. At the first attempt, the match was called off because of a waterlogged pitch, but not before Hereford had incurred expenses of £600 (£4,800 today) for a wasted journey. Attempt number two was scuppered by 5 cm of snow on top of the waterlogging. At the third time of asking, the game went ahead and somehow little Hereford came away from the North-east with a 2–2 draw and a replay at their own Edgar Street ground. Now it was Newcastle's turn to travel in vain, as the Hereford pitch was sodden. Severe frost put paid to the fifth date, but Newcastle decided to stay nearby until the match could be played. They waited all week, cooped up in a local hotel, making do with inadequate training facilities as neither the local RAF station nor the deep mud of Worcester racecourse proved satisfactory. The game finally took place in front of 15,000 Hereford fans on a muddy but playable surface. With ten minutes left, Newcastle were leading by a single goal but the local side equalised and, in extra time, scored the winner. (The giant-killing continued a little longer. In a home tie against West Ham, Hereford drew 0–0 on their heavily sanded, bottomless pitch before losing the replay 3–1 at Upton Park.)

It takes extreme circumstances for international rugby matches to be abandoned – tough breed, rugby players! The England v Scotland fixture at Twickenham even took place in the notorious winter of 1947, the snowiest of the twentieth century.

When the sides met in March the pitch was hard and the bitter cold was accompanied by a second-half blizzard. The Scots, none of whom had taken part in a competitive game for two months, lost 24–5, but their heroics in getting to the match at all deserved credit. Most of the team had travelled to London on Thursday afternoon and endured a 20-hour journey, but two players, who set off later, suffered a worse fate. One spent all Friday, day and night, on a train with no buffet car, the other was stuck in a train that had to be dug out of snowdrifts twice before it reached the border. Both failed to arrive in London until Saturday morning. But at least the game was played. When an equally intrepid Scotland party set out to play Ireland at Lansdowne Road in February 1933, the match had to be postponed, as a snowstorm had left the ground under several centimetres of slushy snow. Unfortunately for the Scots, it also prevented their ship from docking, and they spent 16 hours wallowing offshore in Dublin Bay in blizzard conditions. They were probably very relieved to find the match was off, as few were in any state to play a rugby international when they finally got ashore on the Saturday morning.

More recently, 1987 qualifies as a landmark year. Both the Calcutta Cup match between England and Scotland at Twickenham and the Five Nations fixture between Wales and Ireland at Cardiff Arms Park were postponed on 17 January, not because the pitches were unplayable but in the interests of public safety. Heavy snowfalls across the country five days earlier had brought traffic chaos, with trade and commerce virtually grinding to a halt in some areas. There was little improvement as the weekend approached and, once again, the Scots had travel problems. Some players were trapped in the snowbound Borders, flights from Edinburgh airport were subject to disruption and the police advised travel within mainland Britain only in cases of life and death. Dudley Wood, secretary of the RFU, choosing not to take the Bill Shankly approach to sport, announced the postponement of the England v Scotland fixture, commenting that 'Leisure comes second to life and limb.' It was the first time in 40 years that weather had delayed an international at Twickenham. Given the excellent condition of the pitch, it would have been better, from a rugby point of view, to have played in January, when at least the day was fair. The rescheduled fixture in April took place in torrential rain on soft ground. The slippery ball made handling difficult but England ran out 21–12 winners.

Any lengthy period of snow or frost plays havoc with National Hunt racing. Even an isolated instance of sharp overnight frost can result in racing being called off, as one of the authors discovered during a generally mild February in 1996. Only a mid-morning radio announcement prevented a wasted journey from central London, where it was fair, dry and not particularly chilly, to suburban Sandown, where the card had been abandoned. No risk can be taken, however, with the safety of horses and jockeys and a bad winter can see both confined to stables for weeks at a time. The appalling winter of 1947 saw 51 consecutive days without racing and the postponement, and then abandonment, of the National Hunt Festival at Cheltenham. (The principal races were later run in April.) At the beginning of

March, the jumps at Market Rasen were buried in snowdrifts and there were three metres of frozen snow at Southwell. In 1997, 61 race days had already been lost to weather by 13 January, including a two-week spell before and after New Year. Even the euphemistically named 'all-weather' racing can fall foul of wintry conditions. In very low temperatures, the surface of the all-weather tracks, composed of sand and synthetic fibres, becomes crusty and lumpy, and constant harrowing is needed to break down the clods. In any case, all-weather meetings only provide flat racing. They are a means of keeping the betting industry alive and some horses fit, but cannot replace National Hunt fixtures.

As if the regular snowy season is not enough, winter has a habit of extending into the months of so-called 'spring'. Since the mid-1960s, there have been as many white Easters as white Christmases, and spring snow has influenced numerous events. The winter of 1998 was not harsh – the Premiership did not lose a single game to the elements in that year – yet Easter weekend weather did its best to wreck sport in many parts of the country. While the sun shone bright and warm on the south coast during this middle weekend in April, there was snowman-building in Dyfed and severe flooding on the Rivers Nene and Ouse. The Melrose Sevens and the Yorkshire derby between Barnsley and Sheffield Wednesday took place in snow flurries and biting winds, and a snowstorm disrupted the World Superbike qualifying at Donington Park in the East Midlands. British superstar Carl Fogarty had been lying 14th after two timed sessions but had the opportunity to improve his position in the third. He recorded the fastest time and his best of the weekend only to see the heat cancelled because of snow. Grid positions were declared on the basis of the first two qualifying sessions – Fogarty was not amused. In Scotland, 5 cm of fresh snow fell on the Cairngorms but many skiers were out of luck as the higher latitude runs were closed because of fog and, later, high winds. Sub-zero temperatures were recorded all day at three of the Scottish ski resorts and, with an estimated 30,000 hillwalkers out in the Highlands, the emergency services were advising extreme caution because of the risk of avalanches and white-outs. The annual cross-country races took place at Balmoral on an estate covered in snow and ice. In spite of the conditions, Paula Radcliffe, wearing gloves, managed to knock 15 seconds off the world best time for the 5-mile race. On the busiest day of the point-to-point season, at least half of the meetings were called off. The Scottish boys' championship golf final in Aberdeenshire was constantly interrupted by driving snow and hail, and had to be stopped for lunch after 16 holes, although it was completed in better weather that afternoon. By then, the snowband had moved south to disrupt the Craigmillar Park Open at Edinburgh. This was shortened to 63 holes because of heavy snowfalls.

Why do British sports administrators continue to cram competitions into those dodgy months of March and April? Why have so many sports events traditionally been held in spring when the weather can be especially unpredictable? It is, of course, a cross-over period, when the rugby, National Hunt racing and football seasons reach their climax and flat racing, outdoor tennis, tournament golf and cricket emerge from hibernation: prematurely, in some instances. The Oval was allegedly covered in snow early on Easter Monday, 21 April 1908, when a 60-year-old W.G. Grace played in the final match of his first-class career. April 1989 became infamous for snowy cricket as the first full Sunday of the new season succumbed to the weather. The match between Derbyshire and Northamptonshire on 23 April was snowed off and the game between Worcestershire and Nottinghamshire on 25 April was delayed by snow flurries. In April 1994, Cambridge University v Nottinghamshire was interrupted by a snowfall of several centimetres, and umpire Dickie Bird, in his autobiography, recalls building a snowman beside the stumps when the skies finally cleared. Kevin Mitchell of *The Observer* noted recently that between 1997 and 2001 it snowed three times on the first day of Oxford University's opening match against a first-class county.

Cricket is not the only summer sport to be affected by winter chill. Early April snow and bitter cold in 1968 saw racing abandoned at Nottingham and disruption to the training schedule of the main English prospect for the Two Thousand Guineas. Petingo was not allowed onto the gallops at Newmarket because of the severe weather. A month later he was beaten by the Irish-trained Sir Ivor in the opening Classic of the season. On 26 April 1981, the 50th British hard court tennis championships at Bournemouth were driven inside by a combination of rain, hail, sleet and snow. The men's final began outdoors but was completed on a carpeted indoor court. Strong, bitterly cold winds, reducing temperatures to freezing, badly affected golfers at the Benson & Hedges International Open at St Mellion in Cornwall in April 1991. Woolly hats to the fore, they struggled to cope with conditions on the demanding, exposed course. Among the second-round scores were 80s for Howard Clark and Nick Faldo – recently returned from the Augusta Masters where temperatures had been double those in Britain. There was an 81 for Mark James, 83 for Jose-Maria Canizares and 84 for Ronan Rafferty. James, carding a third-round 88, admitted, 'There were moments when I was so cold I couldn't feel my hands.' His score included 3 eights and 2 sevens. (Ironically, the inaugural Benson & Hedges at the Cornish course had taken place in a heat wave the previous year.)

But if freedom from occasional wintry weather became one of the criteria for summer sports seasons, when would they ever start? Racing at Haydock was abandoned in May 1935 because of snow, and hail stopped play at Wimbledon on 26 June 1980. Probably the most notorious summer snowfalls occurred in early June 1975 when two County Championship cricket matches were abandoned in conditions more reminiscent of January. As usual on these occasions, the stories have been handed down in a variety of forms. According to newspaper reports, more than 3 cm of overnight snow lay on the ground on Monday, 2 June, at Buxton, where Derbyshire were playing host to Lancashire. The game at Bradford between Yorkshire and Leicestershire was also abandoned for the day. Umpire Dickie Bird, perhaps exaggerating slightly, remembers 15 cm lying on the Buxton pitch – maybe it was his faulty metric conversion table! Although it melted away, the wicket was tricky to say the least when play resumed on Tuesday. On a vicious and virtually unplayable surface, Derbyshire were all out for 42.

Wintry weather makes life difficult enough for those sports whose seasons run from September to May. The April day in 1981 that disrupted the Bournemouth tennis championships also put paid to five Football League and three rugby league matches. The Boat Race took place in blinding snow and hailstones in March 1952, in the same month as England played Ireland at Twickenham. There were banks of snow piled high on the touch-lines, groundstaff were still attempting to clear the pitch as the players emerged and the game was played in a blizzard. The Cheltenham Gold Cup had to be temporarily postponed in March 1987 because of a snowstorm, and similar conditions the day before the 1994 Grand National led to an early inspection. The race was run but the going was heavy.

Every spring also brings at least one story of drama and sometimes death amongst hillwalkers. At the end of April 1981, five teenage air cadets got lost on Dartmoor on a training exercise, during what was described as the worst spring weather for a decade, and the worst April snow since 1950. Up to 20 cm fell and the Dartmoor rescue group abandoned an all-night search for the boys after visibility dropped to 5 metres. The cadets were found 24 hours later suffering from exposure and mild hypothermia, having pitched their tent and stayed inside until conditions improved. Had they not done so, the rescue party believed they would have died. In early April 1989, five teenage Guides on a Duke of Edinburgh expedition had to be rescued in a remote part of the Yorkshire Dales after camping out all night in a blizzard. The girls were well equipped for bad weather and had decided to pitch their tent after encountering driving rain, sleet and snow at an altitude of 750 metres above Skipton. They were discovered the following day by the Upper Wharfedale fell rescue team, cold and wet but unharmed.

In addition to legitimate winter games and prematurely staged summer pastimes, there are the plainly eccentric events which are held in spite of the worst the winter can do. Less well known than the Boat Race is the annual competition organised by the Oxford and Cambridge Golfing Society for the President's Putter. It was started as an amateur contest in 1920, has been won by Ted Dexter, ex-England cricket captain, and numerous members of Walker Cup teams and takes place at Rye. There is one very strange fact about the contest, however: it is held in January, and consequently there are often problems with the weather although it has only been cancelled once in over 80 years. In 1960 the final was reduced to 15 holes and in 1963 it was moved from snowbound Rye to Littlestone golf course where it was played in bitter cold and strong easterly winds. In 1982, 1985 and 1987, the latter stages were completed in March after snow halted proceedings in January. In 1993, the final was abandoned after five holes because of high winds and, in 1997, the whole competition was delayed until February as Rye was again under 7 cm of snow. But in 1979 the usual assortment of coloured golf balls, hand warmers and hip flasks could not save the tournament; snow on the course prevented it from even starting.

The Powderhall Handicap Sprint has an even longer history. Held annually in Edinburgh since 1870, it has always been a professional race, the subject of many betting coups in the era of overwhelmingly amateur athletics. Like the President's Putter, one of its claims to fame is that it is run in winter, originally on New Year's Day, but in recent years during the Christmas/New Year holidays. Since 1999, it has taken place at Musselburgh racecourse as part of a holiday race meeting; or at least it would have if normally frost-free Musselburgh hadn't succumbed to the cold in 1999 and 2000. With racing cancelled, the sprint went ahead on a specially prepared track in front of several hundred hardy souls. In 2001, it was third time lucky as both the renamed New Year Sprint, worth £6,000 to the winner, and the National Hunt fixture beat the weather. Although the 110-metre race has been rearranged, it has never been cancelled. It must have been a close call in 1952, however, when the first overseas winner, Eric Cumming, an Australian sheep farmer, had to battle through 5 cm of snow at the Powderhall Greyhound Stadium, from which the race originally derived its name.

The only athletes who pray for heavy snowfalls are winter sports fans. British skiers and snowboarders can seldom get enough of the stuff, as the Scottish mountains provide very unreliable and unpredictable conditions. (Contrary to popular perceptions south of Watford, very few areas of Scotland have significant snowfall.) The Scottish ski industry is entirely dependent on the weather and the type of snow it produces. All five Scottish ski resorts have to work extremely hard to keep runs open, erecting fences everywhere to trap drifting snow. They are helped in this by the wind, a prominent feature of the Scottish climate in the winter months, especially in the mountain areas. But year-to-year conditions vary enormously. The winter of 2001 saw heavy snowfalls and some of the best skiing for a decade whereas a mild winter, such as 2002, makes life very difficult for the skiing industry. The British ski championships were moved out of this country many years ago because of unreliable weather but, ironically, they were cancelled in 2001 due to lack of snow in Austria when they could have been staged in Scotland! Even Alpine resorts have increasingly suffered from erratic snowfall in the last 15 years, with World Cup races scheduled for December having to be postponed because of too little, or occasionally too much, of the essential ingredient. Some of the shortfall can be made up by the use of artificial snow but heavy falls and blizzards usually result in the rescheduling of events. Frustration seems to be the name of the game throughout Europe for those who would like to enjoy winter sports.

SKI SCOTLAND

The five Scottish ski resorts are all within 180 kilometres of the major cities of Edinburgh and Glasgow, with Glencoe and Nevis Range in the west, and Cairngorm (Aviemore), Glenshee and The Lecht in the east. Skiing began at Glenshee and Glencoe in the 1930s while the first chairlift began operating at Cairngorm in 1961. Forty years later, it was replaced by a controversial (but weatherproof!) funicular railway. There are snow-making cannon and piste-grooming machines at several sites, as well as snowboard fun parks to augment the natural facilities of the resorts. The Lecht, situated on both sides of the A939, the Cockbridge to Tomintoul road (which has an unenviable record as one of the most frequently blocked routes in the country) also boasts an artificial ski slope and its own Race Training School. Nordic ski centres and a large number of ski and snowboard schools are also found at the main centres. All this infrastructure is dependent on the arrival and retention of adequate snow of the right type, and the absence of high winds, which can close the chairlifts and put the higher runs out of action: hence the importance of the new Cairngorm railway. If ever a sports industry was at the mercy of the weather, this is it!

Another sport that benefits from a hard winter is curling. According to a correspondent from *The Scotsman* in 1979, 'curling is regarded as a pleasant form of mild insanity linked with weather conditions' – the outdoor form of the sport, that is. It has been almost exclusively an indoor game for nearly a century, taking place on smooth, consistently even sheets of ice. But sometimes the weather is harsh enough to freeze shallow lakes to sufficient depth that matches can be played outdoors, as they would have been before the invention of mechanical freezing. When 21 cm of ice formed in February 1979 on the Lake of Menteith near Callander, some 40 kilometres north of Glasgow, over 3,000 curlers and spectators, including some from Wales and England, converged on the area to take part in the Grand Match, an outdoor competition planned every year by the Royal Caledonian Curling Club but seldom enacted.

Although it looks pretty and benefits some minority leisure activities, wintry weather can be very costly for mainstream sport. There is the inevitable dislocation as fixtures have to be rescheduled and reshuffled into increasingly congested spring weeks, but there is also a financial burden. Newcastle lost a television fee from Sky when their match against Southampton was called off only two hours before kick-off in February 2001. In the same month, Plymouth Argyle were relieved that their match against Rochdale was postponed 24 hours in advance, saving them a wasted journey. The game had already been called off three times and they had twice endured abortive trips north, together with the costs of hotel accommodation. As in the case of the 1987 rugby internationals, sometimes it isn't the state of the pitch that leads to postponement, but difficulties with travel, access roads, parking or spectator facilities. When Scarborough were snowed in and unable to travel south for their Division Three game against Leyton Orient in January 1996, the Orient chairman said the affair had cost the club a lot of money on catering, and that he would be speaking to the Football League on the following Monday. Scarborough had been advised by police not to travel. The match between Aston Villa and Leicester at the end of December 2000 was called off, not because of the pitch but because frozen pipes, car parks and pavements would have prevented spectators from attending.

What seems surprising is how little change there has been in the precarious finances of football clubs in the past 40 years. In the winter of 1963, the Football League made loans available to clubs that were struggling without the cash flow of home gate money. In the winter of 1982, the same financial problems still existed, with several clubs in the lower divisions facing an acute lack of income to cover players' salaries and interest payments on loans. By mid-January that year, 175 games had been postponed because of bad weather and some clubs had not played at home for seven weeks. The *Financial Times* estimated that lost gate money at Christmas alone amounted to £1 million. Although some of this would be recovered when games were eventually played, a proportion would never be recouped as fans struggled to pay for extra tickets within a shortened time scale. In 1997, the loss of two consecutive home fixtures to weather brought with it significant cash-flow problems, even for clubs in Division One. Crystal Palace were quoted as losing £250,000 in home gates over the Christmas period. The Premiership, where finances are more secure, is ironically the division that suffers least from weather disruption. Here, many clubs have been able to invest in undersoil heating but, with installation costs above £250,000 and maintenance of over £20,000 a year, this solution is unavailable to most smaller outfits. The benefits speak for themselves; Arsenal haven't lost a match to a frozen ground since 1985.

The next time snow, frost, ice, bitter winds, sleet or any other combination of winter misery wipes out the sports fixtures, spare a thought for the minnows of league football and the cash-strapped teams, of which there may be many more in the years to come. In an era when it is not unusual for administrators to be called in to save ailing football clubs, a match postponed by the weather can be the last straw. When Greenock Morton's home game against Ross County was called off in late December 2000, the anticipated cash to pay players' wages evaporated and they were thrown back onto payments from the fans' fighting fund. There were pleas for supporters to make extra donations, public rallies on behalf of a club founded in 1874 and even promises of contributions from rivals when Morton fans attended away fixtures. The postponement was also too late to prevent Ross County, based in Dingwall and the most northerly member of the Scottish League, from setting out on a hazardous 500-kilometre round trip. They had travelled as far as Stirling before they heard that the game was off and were angry at the wasted time, money and effort. Perhaps this is yet another occasion for that hackneyed headline: the weather won!

(Postscript: Morton are still surviving – just!)

6

Fog on the Tyne . . . and Other Places

November forms part of the season of mists and mellow fruitfulness, especially the mists that often turn into dense fog. That's what happened on 21 November 1945 when Dynamo Moscow played the third game of their four-match British tour in a London pea-souper. The game was against Arsenal but it was held at Spurs' White Hart Lane ground because Highbury had been requisitioned as an Air Raid Patrol Centre and was not yet fit to host such an important match. It is doubtful if either of these north London venues was really suitable for play that night. Many of the players complained that visibility was so poor that fouls went unseen. Stanley Matthews, guesting for Arsenal on the wing, recalled that he could not see more than a metre or so. Full-back Bernard Joy went to tackle his opposing player only to discover that his identity had changed: Dynamo had taken advantage of a substitute rule for the game and swapped a player without even telling the referee. Some believe that for a while Dynamo actually had 12 men on the pitch. The referee, a Muscovite who had accompanied the team to Britain, used the Russian system of match supervision with the linesmen – as assistant referees were properly called in those days – both operating on the same touch-line while he patrolled the other. Whether they could see sufficiently to indicate offside is debatable and whether the referee could see them, let alone their upraised flags, is equally doubtful. There were many dubious decisions in a game in which Dynamo came back from being 3–1 down to grab a 4–3 victory. If the match officials and players had problems knowing what was going on, then what about the crowd? Roy Peskett, football correspondent for the *Daily Mail,* summed up the match as 'one of the most exciting games 54,000 people have never seen!'

In the often smoke-filled atmosphere of Britain's industrial cities, fog could often worsen into smog. This is less of a problem since the Clean Air Act of 1956 that helped rid city air of most of its smoke content by means of smokeless zones and an encouragement to use smokeless fuels. Prior to that, London fogs in particular were renowned for their thickness and opacity, something to which Sam Bartram can testify. Bartram played in goal for Charlton Athletic 583 times, including one home match at The Valley in the late 1940s against London rivals Chelsea in which the fog rolled in soon after kick-off. In his autobiography, Sam recalled that the referee stopped the game once but restarted it when visibility improved. Sam was untroubled as Charlton gained the ascendancy and most of the play was in Chelsea's half. Then a figure in blue appeared out of the gloom before him, not a Chelsea forward but a policeman who told Sam that the game had been abandoned a quarter of an hour earlier and the rest of the team were enjoying a hot bath.

The infamous London smog of December 1952, in which the chemical-filled air turned a dirty, yellow-grey, lasted for five days. For 60 hours, visibility in parts of the capital was officially reported as less than 20 metres. Whether football referees could see the goals from the halfway line became immaterial as London's transport system descended into chaos. Bus conductors, often wearing masks, walked in front of their vehicles trying to pick out the route. There was little chance of either spectators or players reaching venues and matches were called off. Yet cancelled games were only a small price to pay: elsewhere in London and the Home Counties some 6,000 deaths were attributed to the smog.

FOG, SMOG AND MIST: CAN YOU SEE THE DIFFERENCE?

Both mist and fog occur when the temperature of air at ground level drops to the point where condensation begins. The first water droplets to form are small and far apart, thus producing mist. Later, if the droplets increase in both size and number, the mist thickens into fog. In weather forecasts for the public the word 'fog' is often used only when visibility is expected to be less than 180 metres. When air contains a large amount of smoke or other pollutants fog forms on the many tiny particles earlier and easier than if the air was clean, forming smog, a term derived from a combination of 'smoke' and 'fog'.

Meteorologists define fog as occurring when horizontal visibility is less than a kilometre. In no sport would this be regarded as a serious hindrance. At 200 metres visibility, however, 'fog' officially becomes 'thick fog' and can be a major problem in motor sports and golf. In 1978 it held up the World Matchplay Championships at Wentworth and delayed the scheduled Sunday final until the Monday, when Japan's Isao Aoki beat New Zealander Simon Owen three and two. 'Dense fog', with only 50 metres visibility, is sufficient to stop most outdoor sports. But not at Derby racecourse in 1889 where the stewards insisted on the meeting going ahead although visibility was said to be less than 45 metres. Policemen lined the course to help jockeys find their way and blew whistles to warn the crowd of the approaching horses but the riders still became lost, and, at full racing speed, collided with a set of hoardings. Fortunately, the only injuries were cuts and bruises. Ironically, a century later, fog led to the abandonment of only the third race meeting on the euphemistically titled 'all-weather' course at Southwell.

Some players seem able to cope with, or are at least willing to risk, foggy conditions. On St Andrew's Day 1907, members of Royal Liverpool Golf Club turned up at Hoylake for their monthly medal competition only to find it cancelled because of the poor visibility. However, one of them, John Ball, eight times amateur champion and Open champion in 1890, decided to accept a challenge to go round in fewer than 90 shots. Playing with a black ball on a course that he knew intimately – he had won 94 medals there – he shot nine under his target.

Although a ball or two might be lost, golfing in the fog is not too hazardous for the individual player. This is not the case in motor sport where lives can be at risk. But drivers, men who push themselves and their cars to the limit, can be crazy. When the 1989 Motaquip British Rallycross Grand Prix was abandoned due to fog blanketing the Brands Hatch circuit, the drivers opted to decide the race by individual time trials over one and a half laps, despite the fact that corners came up at frightening speed and memory more than vision was the significant factor in getting round them. Sometimes playing in the fog can be more of a comedy than a tense drama. After struggling through the gloom to reach the Richmond Athletic Ground on the afternoon of 22 December 1962, neither the Richmond nor the Harlequins team were sure that there was sufficient visibility for their rugby game to go ahead. Five of the waiting players decided that play would not be possible and headed home. The referee, Air Marshal Sir G.A. Walker, did not share their opinion and substitutes had to be found. While they were changing, the fog thickened but the game kicked off. Spectators could not see anything except when play neared the touch-line. Whether the players and referee were any better placed is debatable. Eventually, after half an hour of hide-and-seek in the murk, the referee announced that the game was a 0–0 draw. Allegedly he then added, 'For the sake of honour we are playing on.' There is no official record of what happened after that.

Fog is a major problem for sailors, as drastically reduced visibility at sea can cause spatial disorientation. Fortunately miniaturisation of electronic navigation equipment has come to their aid. Radar can penetrate the gloom; an echo sounder can supply information on the depth of water under the boat and enable the yachtsman to follow seabed contours back to the shore; and position indicators making use of satellites can place a boat within 200 metres, even less if special facilities are paid for. Certainly Thomas Lipton, the tea magnate, could have used one of these devices back in 1903 when his yacht *Shamrock III*, challenging for the America's Cup, got lost in fog off the New York coast and failed to finish, thus conceding defeat to *Reliance*, the defending boat. Yet instruments can go blank and modern yachtsmen, like their non-electronic assisted predecessors, may have to rely on the old-fashioned chart, magnetic compass, pencil and ruler as well as experience and skill.

Some events are called off in fog because of safety reasons. Obviously those involving speeding drivers, motorbike riders or cyclists often fall victim to reduced visibility. The Tourist Trophy motorbike races on the Isle of Man are dangerous enough at any time, accounting for nearly 200 deaths since they began in 1907, and officials are reluctant to add to the risks. In 2000 they delayed the lightweight TT for four and a half hours while the riders waited for the mist to lift from the mountain section of the course. The race, shortened to three laps, was won by 48-year-old Joey Dunlop, who collected his 25th TT victory on the circuit. In April of the same year, thick early morning fog in Northamptonshire caused the warm-up for the British Grand Prix at Silverstone to be postponed, not because the drivers were unable to see but due to the grounding of the helicopters used to rush any injured personnel to hospital. Cricketers have been taken off when the safety of fielders has been jeopardised by them not being aware of where the ball was going, or, even more dangerously, where it was coming from. In August 1959 there was a 40-minute interruption for fog at Edgbaston where Warwickshire were playing the touring New Zealanders. Suspensions through fog are more common at seaside cricket grounds, such as Swansea and Scarborough, where mist and sea frets can drift in from the ocean. Other sports called off because poor visibility compromises safety include fishing matches, at which anglers might fall unseen into rivers and canals, and road races where traffic could inadvertently hit runners, even those that had not lost their way.

Some events are abandoned because the adjudicators cannot see what is happening. The standard test in the football codes is whether the referee – the sole arbiter in matters of visibility – can see both goals from the halfway line. In November 1989 (that month again) the referee in the rugby league match between Hull Kingston Rovers and Workington abandoned the game when he decided that he could not follow the play. This was a technically correct decision, but perhaps an irrational one as there were only three minutes left to be played and Hull were leading 34–0. The Rugby League authorities sensibly did not insist on a replay and awarded Hull two points for the win. Judging in the traditional rural sport of coursing – in which greyhounds chase live hares – can be affected by poor visibility as the dogs are assessed by how they turn their quarry, not just on the kill itself. Racing can be cancelled if the judge cannot see the finishing line or if the stewards cannot see what is happening 'out in the country'. Bookmakers, always with an eye on their satchels, have been known to take bets on whether a meeting will be postponed, as they did at Lingfield in 1987 when freezing fog lingered around the course. On New Year's Day 1996 the entire meeting at Southwell was called off and that at Exeter abandoned after the fifth race. Exeter, which borders Dartmoor, is more susceptible than most other courses to mist and fog. More recently, on 27 March 2000, the opening day of the point-to-point season, fog came down after the fourth race at the Vale of Clettwr meeting and ruled out the remainder of the card. It turned out to be the last race of the year as the rest were cancelled because of the foot-and-mouth outbreak. According to statistics in the *Racing Calendar*, National Hunt meetings are almost six times as likely to be cancelled as those on the flat. This simply reflects the dominance of jump racing at times of the year when fog is more prevalent. The current Jockey Club rules state that, at the beginning of a race, the starter must have the recall flagman in sight and that, at the end, the judge has to be able to see at least to the furlong (200 metres) marker so that he can identify the horses as they approach the finish. How one jockey might have wished that this applied at Cheltenham in 1841 when he failed to identify the correct winning post in the fog and eased up, only to be overtaken before the real finishing line.

Sometimes fog is localised and patchy, affecting only one sports event, but occasionally it is widespread and causes severe disruption to fixture lists, particularly when associated with other bad weather. On Saturday, 22 December 1962 it played havoc with the football card: 18 matches never began and a further 8 games were abandoned after play had started. This was a disastrous day for the sports fan as two National Hunt meetings were also 'fogged off'. Another Saturday to be affected was 5 December 1976 when frost and fog combined to wreck the second attempt to play most of the first round of the rugby knock-out cup. Of eight matches due to be played, only two were completed. Fylde v Solihull started but had to be called off soon after half-time.

Possibly the major effect of fog on sport is that it makes travelling difficult. Eddie Hapgood, trying to get back to London from his army unit, missed playing in the Arsenal v Dynamo Moscow game because he was stranded in Brussels with his flight grounded. In 1990 in the PES English Counties Netball League, top of the table Middlesex were thrashed 51–27 by Surrey after the coach and two leading players were delayed on their return from holiday in Italy and arrived in time only to watch the last quarter. Not only air travellers can be late. Two golfers in the Amateur Championship at Sandwich in 1937 fell foul of poor visibility at sea. Both Brigadier-General Critchley and a Burmese amateur were on the *Queen Mary* sailing from New York to Southampton, a journey that was delayed by fog. The golfer from Burma opted to take a car from Southampton and was four hours late for his tee-off time so, despite travelling more than halfway round the world, he was not allowed to start. The Brigadier chartered a plane to Sandwich and circled over the clubhouse to let officials know that he was on his way, but he arrived six minutes late and he too was struck out.

Not all participants travelling – or attempting to travel – in fog are as unlucky. The 1989 World Cup of Golf at Las Brisas in Spain almost started without several of the competitors, who were stranded in a shrouded Heathrow. They all got there, however, including the eventual winning Australian team of Peter Fowler and Wayne Grady. In January 1997, jockey Adrian Maguire was scheduled to fly to Leopardstown to partner Mulligan in the Baileys Arkle Perpetual Challenge Cup. He was delayed by thick fog at Biggin Hill and had to drive to Farnborough for another flight to Dublin. Along with the horse's owners, he was driven with an escort of Garda motorcyclists to the racecourse and arrived just in time. The group appreciated the assistance as Mulligan won the race, beating the favourite, Danoli.

Less success came the way of Ballydoyle trainer Aidan O'Brien in the 1998 Derby. His trio of horses, King of Kings, Saratoga Springs and Second Empire were to be despatched by air to Epsom on the morning of the race, but poor visibility delayed the incoming flight to Shannon. Another plane was made ready at Shannon but it could not land at Farnborough, where O'Brien's horsebox was waiting, as the runway was too short. Southampton was briefly considered but it was discovered that the airport did not possess a horse ramp. Finally Luton was decided upon and arrangements made with Epsom trainer Philip Mitchell to have a horsebox ready. Like Maguire, the Irish contingent received a police escort and they made Epsom by 1 p.m., less than three hours before the race, a much tighter margin than O'Brien had planned for. The late arrival did not help the horses. Both King of Kings and Second Empire were sweating profusely before the off and never featured in the race. The best placed of the trio was Saratoga Springs and he managed only tenth.

Yet another horse to travel but wonder if the trip was really necessary was French Holly, who took on champion hurdler Istabraq at Leopardstown in January 1999. He had a tiring journey. Originally due to be flown to Ireland, he was stranded in fog and sent back to Middleham. Halfway through his supper he was scrambled back into the horsebox to make a dash for a last-minute vacancy on the ferry. He got to the Irish course but was soundly beaten. Two months later, with much less hassle, he again met Istabraq in the Champion Hurdle at Cheltenham and, although beaten again, this time he was a creditable third. Another French connection with reason to curse the fog was the national rugby league team during the 1987 Whitbread Trophy World Cup. They were to play Great Britain at Manchester. Both teams had recently been humiliated by Australia and were determined to begin a renaissance. The weather worked to Britain's advantage. The French squad was supposed to arrive in Manchester on the evening before the game, but poor visibility in the North meant that their plane was diverted to London. They had to board a coach, arriving at their Manchester hotel at two in the morning, hardly ideal preparation for such a crucial game. France lost 52–4.

Others to travel and regret the journey include the supporters of Tottenham Hotspur. On Wednesday, 5 December 1962 *The Scotsman* reported that on the previous day Glasgow had suffered from a dense fog, one of the worst for five years. Presumably Spurs fans did not read 'foreign' newspapers as thousands of them had no qualms about coming up to Ibrox that day where their team was to play Rangers in the second leg of a European Cup-Winners' Cup tie. Conditions had not improved as kick-off approached. Visibility in the early afternoon had been reduced to no more than 9 metres, and, although later improvement enabled the referee to see some 45 metres, the French official assigned to the game had no hesitation in calling it off. This decision was taken less than an hour before kick-off. As it was, trains from the South were running hours late and when Spurs fans arrived in Glasgow, disgruntled and unhappy, they were greeted with the announcement that there would be a special train back to London in less than two hours. Few wanted to return north the following Tuesday, whatever the weather, and Rangers fans congregated on the platforms to take any spare tickets going for the replay. They might have wished that the first game had gone ahead under the foggy cloak of invisibility. At least they would have been spared the sight of their team losing 3–2 and going out of the competition.

THE INVISIBLE MAN . . . AND HIS HORSE

Some participants actually welcome invisibility. Fog does not always stop horse-racing and the unscrupulous can take advantage of the mist. In 1990 Sylvester Carmouche was banned from riding for ten years by the Louisiana stewards after taking a short cut to win a race at Delta Downs. He dropped out near the starting gate, hid in the thick fog and rejoined the pack after a circuit had almost been completed. No similar offence has been detected in Britain, although the shock win of 100–1 outsider Caughoo in the 1947 Grand National led to allegations by Danny McCann, rider of second-placed Lough Conn, that the winner had completed only one circuit of the Aintree course, which was cloaked in dense fog. He insisted that Caughoo had never passed him and must have jumped in at some stage near the end of the final circuit. However, Irish jockey, Aubrey Brabazon, whose mount, the well-fancied Luan Casta, had come to grief at Becher's Brook the first time round maintained that, while he was lying on a stretcher recovering from his fall, he saw Caughoo go past a second time.

WEATHER AT ITS WORST

1

The Penalty is Death

Weather can kill. Golfers struck by lightning, motorcyclists and racing car drivers skidding on wet circuits, climbers subject to exposure and yachtsmen hit by storms at sea have all lost their lives as the climate has revealed its darker side.

One of the most taxing of ocean races, the 970-kilometre Fastnet race runs from Cowes on the Isle of Wight out to the Fastnet Rock off south-west Ireland and back to Plymouth. On Saturday, 11 August 1979, 303 yachts with some 2,700 men and women on board set sail in fair weather down the Solent from the Royal Yacht Squadron at Cowes. For two days the wind held steady, rarely blowing above 25 kph or below 20, but then the clouds darkened and the barometer dropped. When the race started there had been a small depression on the weather charts over Nova Scotia, many hundreds of miles away. Within 48 hours it had crossed the Atlantic and picked up speed. Wind gusts of over 110 kph were reported at British coastal weather stations. Out at sea the winds built up massive waves, some over 12 metres high: one navigator reported that it was like sailing through the Alps!

Ocean yacht racing can be dangerous. Prior to the 1979 Fastnet race, however, less than 40 fatalities had been recorded in over a century of such competitions. Since 1925, when 90 sailors in 9 vessels had contested the first Fastnet race, there had been only two fatalities in this event: an owner swept overboard in 1931 and a Belgian skipper who had died from a heart attack in 1977. Yet the sea was not to be trifled with. The organisers of the race, the Royal Ocean Racing Club, whilst insisting that it was the skippers alone who decided whether or not to race, required boats to be equipped with safety harnesses, life jackets and rafts, fire extinguishers and emergency rations. Crucially, however, they did not have to carry a radio transmitter, only a receiver capable of picking up marine weather forecasts.

It was on these receivers that the crews heard the BBC shipping forecast at 5.50 p.m. on the Monday evening. It warned them of a local gale force 8. In the next forecast, at 15 minutes past midnight on the Tuesday morning, the warning had been upgraded to a severe gale force 9. Four hours later some of the vessels encountered gusts of force 11, well over 100 kph. In such conditions many of the smaller boats could not steer around the vast breaking waves, often coming from different angles and crashing into each other, throwing tons of water onto the helpless vessels. All but the largest ocean-going craft became fragile toys at the mercy of the elements. Over a third of the boats were knocked so far over that their masts paralleled the water and a quarter of them capsized totally, some rolling through 360 degrees. The yacht *Adriane* rolled twice; the first time breaking a mast, half-filling with water, and severely injuring one of the crew, and the second time losing a man overboard. Fearing the consequences of the inevitable third roll, the remaining men took to the life raft, one of 24 crews to abandon ship. Five of these vessels later sank.

Those boats without radio transmitters could not even seek external assistance. Their crews faced the storm alone, often with pounding seas and shrieking wind making verbal communication impossible. The task facing rescuers was immense, with 303 boats taking punishment in an area of around 32,000 square kilometres. Nevertheless lifeboats towed in 9 yachts and escorted another 9 to port; helicopters lifted 74 sailors to safety; and commercial vessels and other yachts picked up a further 62. They were the luckier ones. Six men were swept away when their safety harnesses could not take the strain and broke; nine others drowned or died of hypothermia in the cold water and air, either on board yachts or near life rafts that had capsized; and four more were lost from a trimaran that was accompanying the racing vessels. Three bodies were never recovered. It was the worst disaster in the history of yacht racing.

Almost as horrific was a recent Sydney to Hobart yacht race. Beginning on Boxing Day, this annual event heralds the antipodean post-Christmas festivities. But in 1998 there were no celebrations at the end of the race, only thanks for survival and thoughts for the six men that perished, among them British Olympic yachtsman Glyn Charles. The boats were caught in a savage storm in the Bass Strait, the stretch of sea that divides Tasmania from mainland Australia. Waves topped nine metres and winds reached hurricane force. His six shipmates on the Australian boat *Sword of Orion* could only look on as Glyn's safety harness snapped under the pressure of the wind and the maelstrom of an ocean swallowed him up. His body was never found.

Even near to shore, water and the weather can be deadly partners. In September 1999 eight children were left clinging to safety ropes after a teacher ignored advice that it was too windy and insisted on taking them out in the school's 14-foot dinghy. At least they survived, but another pupil, nine-year-old Elizabeth Bee, was trapped beneath the hull and drowned when the boat rolled on a wave only 70 metres out from the harbour at Portsmouth. Another school trip to run fatally foul of the weather at sea involved a group of teenagers from Southway Comprehensive School in Plymouth. At 10 a.m. on 22 March 1993, the second day of their trip to an outdoor activities centre in Lyme Regis, they set off to paddle by canoe along the coast to Charwell. It was about three kilometres, a distance that, even as beginners, should have taken them no more than three hours. At 1.15 p.m., when they should have been safely on board the centre's bus coming back to base, they were still out at sea – no place at that time of year for novices – fighting for survival in the bitter waters of Lyme Bay. They had nearly made it to Charwell, but geography and the weather intervened. Just before the harbour there is a break in the cliffs where the River Char runs into the sea. It is a notoriously windy spot from where windsurfers have to be rescued every year. The force of the offshore wind there steadily pushed the canoes out to sea and into choppier water. Instead of hugging the coast, as had been intended, the paddlers now faced rough conditions out in the bay. Some of the canoes began to fill with water, then capsize, leaving their occupants clinging to the sides. All the time the wind was blowing them further from the shore. Yet no alarm was raised by the centre manager waiting for them at Charwell. Not till a fisherman found an empty canoe at 2.43 p.m. were the coastguards alerted. Not till 3.15 p.m. did they receive a report that a party of canoeists might be missing. Not till 4.00 p.m. was the first rescue helicopter in the air. By now some of the capsized canoeists had been in the water nearly three hours. They were wearing wetsuits that delayed the effects of the cold but, by the time the first victims were hauled from the water at 5.30 p.m., all 11 in the party were suffering from severe hypothermia and 5 of them were unconscious. A trauma ward was cleared at Weymouth hospital and a team of 50 doctors and nurses drafted in to deal with the emergency. Despite their efforts, four of the teenagers died. Industry self-regulation, with guidelines rather than enforceable rules, cost four young people their lives. Two inexperienced and unqualified instructors were allowed to be in charge of the group; no information was supplied to the local coastguards; and no system of alerting rescue authorities had been implemented. The director of the activities centre – a man more interested in sales than in safety – was later found guilty of manslaughter. But, of course, the real killer was the weather, the notoriously dangerous wind that swept the youngsters out to sea and into the cold murderous water.

Worldwide more than 100 people die from lightning every year. Simply because they are outside, sports participants are more at risk than anyone else from its dangers. Lightning poses a threat to cricketers in storm conditions if rain has not already stopped play. In 1987, at Isleworth in Middlesex, nine cricketers were injured and one killed by side flashes from a lightning strike. Golfers especially can be in danger when their umbrellas and clubs make them the tallest object around, a perilous position as the electric force seeks a route down to earth. Lightning fatalities on British golf courses occurred in August 1973 near Birmingham and May 1976 near Ipswich. One of the more publicised golfing deaths was that of Scottish international footballer, John White, in July 1964. He had just driven off the first tee at the Crews Hill Golf Club at Enfield, Middlesex, when it started to rain. Sheltering under an oak tree he was hit and killed by a single flash of lightning. More recently, in March 2000, British businessman Glyn Riley, head of a South African subsidiary of Rolls Royce, was killed instantly when struck by lightning during a storm on a golf course in Johannesburg. He was among a group of eight taking part in a veterans' championship. They had taken refuge under a wooden shelter but the lightning bolt travelled through water on the ground.

ADVICE OFFERED IN THE GOLFERS' HANDBOOK FOR PERSONAL SAFETY DURING THUNDERSTORMS

Do not go out of doors or remain out during thunderstorms unless it is necessary. Stay inside a building where it is dry, preferably away from fireplaces, stoves, and other metal objects.

If there is any choice of shelter, choose in the following order:

Large metal or metal-frame buildings.

Dwellings or other buildings which are protected against lightning.

Large unprotected buildings.

Small unprotected buildings.

If remaining out of doors is unavoidable, keep away from:

Small sheds and shelters if in an exposed location.

Isolated trees.

Wire fences.

Hilltops and wide open spaces.

Seek shelter in:

A cave.

A depression in the ground.

A deep valley.

The foot of a steep or overhanging cliff.

Dense woods.

A grove of trees.

Note: raising golf clubs or umbrellas above the head is dangerous.

In April 1948 two British soldiers, Bertram Boardley of the Royal Artillery and Kenneth Hall of the Royal Armoured Corps, were killed in action: not in a shooting match but a football one. It was in the replay of the Army Cup final between the Royal Armoured Corps, Bovington and No. 121 Training Regiment of the Royal Artillery, Oswestry. The first match had ended in a scoreless draw, but in the replay Oswestry had managed to find the net twice and held their lead well into the second half. Then forked lightning struck. Eight players, including the unfortunate Boardley and Hall, and the referee, a Captain Green, collapsed. Green's rubberised footwear probably saved his life but he was still detained in hospital overnight, as were two players and two spectators. The game was abandoned and the teams shared the trophy.

Lightning *can* strike twice. Royal Ascot race meetings suffered fatalities from lightning in 1930 and again in 1955. The thunderstorm that hit the second day of the Royal meeting in June 1930 flooded the racecourse and caused racing to be abandoned for the day, the first time this had occurred in over 200 years. A bookmaker sheltering in the Tattersalls enclosure was struck by lightning and killed instantly. Elsewhere there was widespread panic. This was captured by the 'Woman Correspondent' of the *Daily Herald* who noted that:

> In some of the tents women fainted. Four collapsed together in one small tent. Word spread that a man had been killed by lightning in Tattersalls. The real terror showed then. Men went pale. Women gazed on the storm with fascinated horror, deadly white. Gradually the density of the rain became less. The air became lighter, the thunder less like the crash of a battery of guns. The frigid stare of terror, incongruous in painted eyes, melted. Women pinned up their frocks in the manner of charwomen and trooped out. It was almost impossible to find their motor cars. Once-smart men and women jostled, ran, searched. Those who set off down the tunnel to the station found two feet of water. They went across the field ankle deep in mud. Their clothes? They were past caring for them.

A quarter of a century later lightning struck again. In 1955 Royal Ascot was postponed from June because of a rail strike and the Gold Cup was held on 14 July. There had been stifling temperatures all week and the weather finally broke that afternoon when a storm swept the course. Thousands ran to shelter from the deluge as a bolt of lightning hit the metal railing around the cheap enclosure and ripped along it in a series of blue sparks. The force of the strike carved a swathe through the crowd on the Heath side and threw over a hundred spectators into the air or flat onto their backs. Two of them died. Forty-nine people suffered injuries including Ronald Batt, whose pregnant wife Barbara was one of those killed by the flash. When the last of the injured had been taken away, her twisted and charred umbrella remained where she had dropped it. The 4.20 race was delayed until 5 o'clock but, when news of the deaths came through, racing was abandoned with two events still to be run.

No matter how skilled the participant, the weather can still have the final word. Although only 20, Alan McNish was an experienced race driver. Aggressive and ambitious, he had begun in karts at the age of 11, gained success at junior racing grades, was a test driver for McLaren, and had been voted National Driver of the Year by *Autosport* magazine. He was clearly one of the outstanding talents in Formula 3000, the level immediately below Formula One. On 22 April 1990, he was competing in the first Formula 3000 event of the season at Donington, the East Midlands racing circuit. It was raining heavily during the warm-up lap on which, uncharacteristically, McNish stalled. This meant that he had to start from the pit lane, in last place. However, the rain eased and the track began to dry. McNish took the opportunity to change quickly from wet-weather to slick tyres. Along with his skill, this enabled him to move up quickly through the field, all of which were on the increasingly unsuitable wet-weather tyres. Towards the end of lap 5 of the 50-lap race, he had passed 10 drivers and was in 16h place. His tyres allowed him to come up fast on those in front, possibly too fast in this instance for, as he closed up behind the Jordan of Emanuele Naspetti at Copse Corner, their wheels touched. Travelling at 240 kph, McNish's Lola cartwheeled into the air and flew over the safety barrier. After three days in a coma, McNish awoke to the news that his car had killed a spectator. He contemplated quitting the sport but the widow of the dead man encouraged him to continue racing. In March 2002, he made his Formula One debut for Toyota at Melbourne.

Like motor racing, climbing mountains is always a risky pursuit, even more so in bad weather. Prior to Michael Matthews' attempt to climb Mount Everest, 161 mountaineers had lost their lives on its slopes. On 13 May 1999, he became the 162nd. Separated from his guide, the 21-year-old London city trader died alone in a snowstorm at the summit of Everest. He was part of an organised leisure climbing tour, one of several commercial expeditions earning large fees by taking amateur climbers into the Himalayas. OTT, the company with whom he climbed, charged $40,000. However, they had an excellent safety record and had never lost anyone on the mountains. They insisted that all the climbers had to prove themselves by conquering high peaks in South America before being accepted on the Everest venture. The expedition took its time and spent days acclimatising at different altitudes. The latter stages required the use of oxygen and there were allegations that the valves on the bottles were not compatible with other equipment. Whether this affected Michael is a matter of conjecture. Towards the end of the climb he was undoubtedly tired and slower than the other climbers, but he was determined to achieve his ambition of standing on top of the world. As those of the party who had reached the peak came down, they passed him on the way up. At the start of the push for the top the weather conditions had looked good with some wind but a clear sky. By the time Michael was within striking distance, however, the weather had worsened. An accompanying Sherpa urged him to turn round but he and his guide, Mike Smith, pressed on. They made the top at midday, roughly three hours behind the others. Temperatures were minus 40°C, the wind was reaching speeds of 160 kph, and it was snowing. In the white-out conditions Michael and his guide became separated and Michael was never seen again. He had been the youngest Briton to climb Everest but, like one in seven of those who reach the summit, he had died on the mountain, another victim of appalling weather.

Mountains do not have to be high to be killers. In bad weather Scotland's hills can be just as treacherous as the Himalayas. As this chapter was being written, two climbers perished on Beinn Dearg, a 1,150-metre peak in Wester Ross. Caught in an avalanche, they had no chance. In order to recover the bodies, the Dundonnel mountain rescue team had to wade waist deep in snow through driving winds of up to 130 kph. Each year between 35 and 45 people die on Scottish hills and mountains. The annual toll of the tired, unwary and unfortunate is greatest on the forbidding walls flanking the eight kilometres of Glencoe. Some who die are brave and unlucky; others are tragically stupid. Most are victims of the weather that, especially in the winter, can overwhelm a crag and its combatants in just a few minutes.

Even apparently far less risky recreational pursuits can be dangerous when the weather is not treated with respect. In March 1964, 24 young walkers set off on the annual Four Inns expedition, a 72-kilometre hike in the Peak District. They began in light wind and drizzle with temperatures around 4–7°C, but the rain became heavy and the wind intensified. Their clothing became wet and lost its insulating power. Wind chill lowered their body temperatures to dangerously low levels. Three of the party died and five more were in a state of exhaustion when rescued. More recently, in October 2000, two other youngsters, Rochelle Cauvet and Hannah Black, died on an activity holiday organised by their school in the Yorkshire Dales. At Stainforth, near Settle, the girls and a few other pupils had waded into the beck and were walking along it through the village. A downpour in the previous 24 hours had left the stream running fast and turbulent, but at first slipping over and getting drenched was fun. Laughter turned to screams as Rochelle was spun round by the swirling waters, could not put her feet back on the river bed and, despite the efforts of teacher Andy Miller, was swept from his grasp to disappear round a bend in the river. He watched in horror as Hannah also lost her footing and was washed downstream. The stream had become an uncontrollable torrent and nothing could be done to save the girls. No one had checked with the manager of the local youth hostel where the party was staying for flood warnings or even a weather forecast.

It is tragic to be killed taking part in sport or leisure activities but there are also risks in travelling during bad weather. One of the saddest incidents in British sporting history involved the 'Busby Babes', the young Manchester United football team put together by manager Matt Busby, and virtually wiped out at Munich-Riem airport on Thursday, 6 February 1958 when their plane crashed taking off in a snowstorm. United, the first English team to accept the challenge of playing in a European competition, were returning from a quarter-final European Cup tie against Red Star Belgrade. The high-scoring game had ended 3–3, a result sufficient to put the Babes into the semi-final following their 2–1 win in the first leg at Old Trafford. Busby was keen to get back from Yugoslavia quickly, as they were due to play Wolves, the leaders of the English First Division, the following Saturday in what many thought would be a crucial match in determining the championship. In the previous round United had been severely delayed in returning from a game in Prague and had got back for their fixture against Birmingham with only a few hours to spare. To lessen the risk of delay this time, United had chartered their own plane for the round trip to Belgrade, a twin-propeller Elizabethan operated by British European Airways. It had been snowing during the game in the Yugoslav capital. Conditions had worsened by morning and when the plane landed at Munich for a brief refuelling stop it was snowing heavily. Almost exactly a year before, following a cold snap, United had arrived at Madrid airport for their flight home only to discover the runway covered in snow and slush. Doubts about the safety of a take-off were finally eased after the players helped to clear the runway. This time no one sought their assistance. The first attempt to take off for Manchester was aborted when the plane was halfway down the tarmac. The second effort met the same fate.

WEATHERBEATEN

The plane's engine pressure was surging too much, a common problem in that type of aircraft. On a runway made treacherous by snow and ice and with a plane whose pilot had failed to check the wings for ice, the third attempt proved disastrous, as the take-off speed could not be maintained. The undercarriage lifted only slightly and the plane smashed through a fence, shot across a road and hit a house. Twenty-one people lay dead or dying in the wreckage: eleven players, eight football writers, the travel agent who had organised the trip and one fan. Two others died later in hospital, one of them Kenneth Rayment, the co-pilot who had been flying the plane, the other Duncan Edwards, just 21 but already with 18 international caps. After the second failed attempt to take off, Edwards had sent a telegram to his landlady in Manchester saying, 'All flights cancelled. Flying home tomorrow.' If only.

8

An Ill Wind: Health Hazards in Sport

Blowing Hot and Cold

The marathon is a long race. That of the London Olympics in 1908 was slightly longer than the runners had anticipated as, at the request of Princess Mary, the start was shifted a few hundred yards nearer to Windsor Castle to afford the assembled Royal children a better view. Those additional yards did for Pietri Dorando, the diminutive Italian athlete who was first into the White City stadium, where the race was to finish. He had not followed the initial fast pace set by South African, Charles Hefferon, but waited till the 24th mile to take the lead. Nevertheless, he was in desperate straits when he reached the stadium. The majority of the 56 starters never even got that far and only 27 of them actually finished the race. Exhausted by running over 26 miles on a warm and humid day with, according to contemporary reports, 'very little air' around, Dorando collapsed on the track, was unsure which direction to run in, and was eventually helped over the line by sympathetic officials, only to be disqualified for receiving such assistance. Queen Alexandra, mother of Mary, later presented him with a replica of the winner's trophy – no gold medals in those days – to commemorate his gallant effort. Dorando was not unfit; he was a victim of heat stroke.

In Britain optimum conditions for comfort are generally accepted by sports medicine experts as being around 15°C and 60 per cent relative humidity. Slight variations to these figures might take sportspersons out of their 'comfort zone' but major changes can be hazardous. Prolonged exposure to environmental heat, for example, can result in heat cramps, heat exhaustion or potentially fatal heat stroke. Heat cramps are painful muscular contractions caused by electrolyte imbalances. Exertion-induced heat exhaustion produces fatigue, drowsiness and vomiting. Much worse than heat exhaustion is heat stroke, where sufferers show signs of mental confusion, loss of motor control and may lapse into unconsciousness or even die. Dorando's experience was a classic example of this: his legs were tying themselves in knots and his brain did not know which way to run.

Almost 90 years after Dorando's brave attempt, England cricketer Chris Lewis was more foolhardy than courageous. He joined the growing fad for sportsmen to have their heads shaved, but then went out to field in a hot West Indian sun without a hat or cap for protection and collapsed with heat stroke. Back in Britain, in July 1976, umpire 'Dusty' Rhodes had to retire with heat exhaustion before lunch at Worcester and 25 years later another umpire, Jeff Evans, officiating in his maiden season of first-class cricket, collapsed at square leg with heat stroke, a result of standing all day without a hat. Those out on the cricket field for long hours are perhaps more at risk from the sun than other sportsmen, but in the summer of 1949 champion jockey Gordon Richards suffered sunstroke at Ascot. Although he was forced to miss over a fortnight of racing, the ill effects were not long lasting and on his return he rode 33 winners in 15 days from 57 mounts and became champion jockey again, for the 22nd time.

Not only summer sportsmen can suffer from the sun. On 1 September 1906, the first Saturday of the Football League season, Manchester City played Arsenal in a temperature of around 35°C. At half-time the home side were down to eight men, as the heat had adversely affected three players. One returned, but three others collapsed with sunstroke in the second half. Arsenal, perhaps more used to sunny southern climes, were not distressed to the same extent and won 4–1. Elsewhere, several other matches became tests of endurance, though sensibly some games were postponed until the late afternoon to avoid playing during the hottest part of the day.

Fluid replacement to prevent dehydration is the key to avoiding heat exhaustion or worse. On their recent tour of India, English batsmen found that they were losing up to two litres of fluid an hour. This was not surprising as they wore padding, helmets, gloves, chest and side guards, carried a 1.5 kg bat, and, perhaps more surprising, occasionally stayed at the crease for over sixty minutes. Footballers in hot climates have lost as much as 5 kg in weight during a match through sweating, but in football there is only one official stoppage – at half-time – when players can take on liquid. Fortunately, during the 1994 World Cup in the United States, and again in South Korea and Japan in 2002, it was also allowed in injury breaks. Golfers playing in humid conditions at the Caltex Singapore Masters in February 2002 – economically, if not geographically, part of the European PGA tour – were advised to drink at least three litres of liquid per round and bottled water was made available at every tee. The marathon, as Dorando found out, is a major test of stamina and hydration. Unaccountably, until 1977 the rule for internationally sanctioned marathons did not allow for any fluid intake till 11 kilometres had been run. This was then reduced to a prohibition on drinking over the first five kilometres, a distance regarded as one at which fluid replacement becomes vital. Sanity has now prevailed and most marathons have water stations virtually every kilometre.

Sport in extreme cold can also be dangerous. Water freezes at 0°C and, as 70 per cent of the human body is water, participants in any activity at sub-zero temperatures run the risk of frostbite. Mountaineers, walkers and skiers in particular have to be careful to avoid it. When the skin temperature drops below freezing, ice crystals begin to form in the tissue, blood vessels freeze, and the skin itself becomes hard and putty-coloured. The lack of blood circulation in the affected areas can produce severe tissue damage. Exposed parts of the body, such as the nose, lips and ears are particularly vulnerable, as are toes and fingers. The danger is accentuated in conditions of frost and wind (or frost and rapid motion, as in downhill skiing). Worse than frostbite is hypothermia, where excessive chilling leads to rapid and progressive mental and physical collapse. This is generally caused by exposure to cold, often aggravated by wet and windy conditions, especially when a person is already suffering from physical exhaustion and lack of food. Victims have a weak pulse, become irrational, slow to respond and have speech and sight difficulties. It can prove fatal if untreated as stranded climbers, disoriented hill walkers and yachtsmen lost overboard have found to their cost.

Water is far more effective – and hence more dangerous – than air in carrying heat away from the body. If water temperature is around 15°C survival time is between four to six hours; freezing water can bring death within 15 minutes. Yet some hardened individuals actually choose to challenge the cold seas. In 1960 Channel swimmer, Florence Chadwick, attempted to swim the Irish Sea at a time when its temperature was about 13°C. After seven hours (and two-thirds of the distance) her accompanying crew dragged her from the water. She had become light-headed, her sight had blurred and she was in intense pain. Her body temperature was found to be 32°C, about 7°C below normal and one of the lowest ever recorded for a long-distance swimmer.

Frostbite should be treated very carefully, as skin cells can be destroyed during the thawing process. Attempting to warm frostbitten parts of the body near a fire or by rubbing – which might appear the obvious treatment – will almost certainly aggravate any damage. Superficially frostbitten areas can be heated by placing them in warm water or against the abdomen or under the armpits of a companion – here's where you find out who your friends are! For more severe frostbite the recommended treatment is to cover the victim's trunk with warm water to encourage the opening up of blood vessels in the affected areas. These can then be gradually warmed by water directly, but with great care as too high a water temperature can kill the skin cells and lead to the loss of an affected limb.

Some of the temperature-induced health problems can be reduced by acclimatisation in which training takes place in a similar environment to that of the scheduled competition. Normally this is achieved by physically moving to a comparable site, but a growing alternative, at least for high temperature acclimatisation, is to use a 'hot box'. Here the athlete exercises under controlled heat and humidity to gain tolerance to unaccustomed conditions. Successful heat acclimatisation results in a decrease in the pulse rate, improved blood flow and an increase in sweating which is the body's thermostatic control. Don Thompson, gold medallist in the 50-kilometre walk at the Rome Olympics, successfully acclimatised himself to the Italian race conditions by training for the event on a treadmill in his steam-filled bathroom back in Britain. The chief physiologist at the British Olympic Medical Centre, Richard Godfrey, estimated that proper acclimatisation of the England World Cup footballers for the summer heat of Japan and Korea would take 10–14 days, during which time the players exposed themselves to the conditions for only 90 minutes a day, whilst, of course, ensuring that they were protected against the sun.

FOUR-LEGGED FRIENDS

Animals can also suffer from the weather while participating in sport. Despite the availability of cushioned horseshoes, trainers are reluctant to run their horses on hard ground as firm going can jolt and damage a horse's shoulder, often affecting its racing career. Rain, too, brings its problems. In wet weather horses can contract mud fever, a form of skin infection that thrives in damp conditions. The risks of hard ground and rain are both increased where horses have to jump obstacles. At the Badminton Horse trials in 1992 it had been dry in the week of the event and many people had walked the course, which further hardened the ground. Then it rained and the water was unable to drain through the compacted layer, which then became slippery. Farriers can provide specialist horseshoes which help to prevent slipping in heavy mud, but they cannot cope with all circumstances. Mark Todd's Face the Music slithered on take-off at a fence and broke a leg. In a classic case of 'shutting the stable door' a decision was made that at all future Badmintons the mass of spectators would be prevented from trampling areas near the obstacles. In these days of global sport with British-based horses competing in racing, show jumping and eventing in the United States, Japan and Australia, acclimatisation is another issue that has to be faced. If not properly acclimatised a horse's respiratory system can suffer.

Covering Up

In his book *Association Football*, Ernest Needham of Sheffield United recalled a match at the end of the nineteenth century against Aston Villa in which the bitterly cold wind and sleet numbed muscle and brain, and men on both sides were carried away for hot baths. Needham himself left the field half an hour before the finish of the game and claimed that this probably saved his life. Some of the Villa players, however, had a unique solution to the problem and played in greatcoats, one of them even carrying an umbrella. Since those days protective clothing has come a long way and all sorts of waterproof clothes and footwear have been designed to help participants and spectators cope with bad weather conditions.

Loss of body heat can be a major hazard faced by athletes in cold and windy conditions. When wind blows through a garment that is not wind-proofed, it destroys the warm layer of air next to the skin. Even wind-proofed clothes are not necessarily insulated and extra layers of thick, loose-knit jumpers or fleecy garments are often worn to offset the wind-chill factor. Hunt followers have always appreciated that the stirrup cup is not enough to keep the elements at bay and have worn tightly woven clothing to minimise heat loss, but their desire for elegance has reduced their ability to counteract the wind and rain. Not so yachtsmen or climbers who, with their lives on the line rather than the fox's, have preferred practicality to looks or tradition. When the Southern Ocean water temperature dropped to 3°C Lisa McDonald, a yachtswoman in the Volvo Ocean race, opted for three pairs of fleece socks, inside Gore-tex socks, inside boots as well as three layers of thermal clothing under her oilskins.

Most of the Fastnet race victims in 1979 died through exposure to the cold water and wind because they were poorly clothed for such elements. The ensuing years have seen dramatic improvements in clothing designed for sailing which keep the wearer dry and at the right temperature. Until the introduction of new materials, the problem with waterproof outfits was that the more features that were added, the heavier they became, ultimately resembling a suit of armour: though thankfully more buoyant! The challenge of events such as round the world races led to substantial improvements in offshore clothing, especially the development of lightweight, vapour-permeable material which allowed for easier movement but also let perspiration escape without letting rain or seawater in.

Keen golfers, anxious to play all year round, have to face wet, windy and cold conditions. They too have benefited from the development of new waterproof materials and an ever-expanding range of bad-weather wear. One bonus is that the new clothing is quieter to wear or, as put by the *PGA Directory of Golf,* 'Gone are the days when playing in a rain suit sounded like a ferret trying to escape from a bag of crisps.'

In hot weather, clothing needs to be light and porous but with enough cover to provide protection from the sun. Marathon participants would be best advised to wear a proper running vest which allows sweat to evaporate and cool the skin, rather than a charity T-shirt; otherwise the heat of the sun and of the microclimate generated by a packed mass of runners can lead to overcooking of the body's cooling system. Where money is no object, specialised equipment can be bought to combat the heat. In the 1984 Formula One Grand Prix in Dallas, cockpit temperatures were as high as 60°C and race winner Keke Rosberg wore a refrigerated skullcap, reputed to have cost £2,500 (£5,000 today). It almost goes without saying that the head should be covered when playing sport out in the sun, but on the recent tour of India the England cricket coach, Duncan Fletcher, had to insist that the players wore caps at all times when fielding.

Cancer publicity has persuaded cricketers to wear all colours of sun block – which also has the advantage that it can be used surreptitiously to shine the ball – but one hazardous aspect of the summer weather only recently appreciated by sportsmen is the effect of ultra-violet light on the eyes. Not till 1996 did Darren Gough become the first England bowler to perform while wearing sunglasses, though wraparound sun specs are now a permanent feature in the cricketer's equipment. The danger of not wearing them was brought home in early 2002 when recently retired umpire

Dickie Bird had to have an operation to restore his sight after the onset of blindness, caused by spending long hours in bright light with his eyes unprotected.

Slip Sliding Away

Then there are the injuries – torn ligaments, damaged muscles and broken limbs – caused by slips, falls and skids on damp or frozen surfaces. Summer showers can make cricket grounds and tennis courts, especially grass ones, skiddy. In 1952 off-spinner Jim Laker slipped while fielding on wet ground at Old Trafford in the third Test against India and dropped out of the attack, after bowling only two overs. More recently, Greg Rusedski, Britain's Canadian tennis star, damaged ligaments in his left ankle by slipping on a wet court in the Stella Artois tournament at Queen's Club in the lead-up event to Wimbledon 1998 and ruined his chances for the Grand Slam event. Ranked number four in the world at the time, he was desperate to compete in the Championship to show how much it and the fans meant to him. In the ten days remaining before Wimbledon he undertook secret rehabilitation. So secret, in fact, that not even his coach Tony Pickard knew where he was, a matter that led to Pickard's resignation. Ironically, Rusedski had hoped that the rain would intervene and allow him another day to recuperate from his injury before playing a first-round match against Australian Mark Draper. However, the rain did not come until he was 5–4 down in the third set, having secured one set thanks to his booming serve, measured at 215 kph, the fastest in the tournament that year. He limped off out of the rain and out of the tournament, his injury aggravated by the attempt to please his fans.

Footballers can generally cope with mud and wet grass; it is icy pitches that bring most grief, both to their bodies and their skills. They really cannot be expected to perform on surfaces more suited to Torvill and Dean. Although 29 matches were postponed in England on an icy 11 December 1976, others went ahead, but five of these were eventually abandoned because of the treacherous conditions, two of them after players were injured. In an FA Cup match at Colchester, Brentford's Terry Johnson emerged from a slide and collision with John Williams nursing a suspected fractured arm. Further north Newcastle manager Gordon Lee insisted on Friday that their Saturday game at St James' Park against Ipswich should not take place but was overruled by the referee. However, the situation deteriorated and after Ipswich centre-forward Paul Mariner, valued at £220,000 (£950,000 in today's prices) had fallen and been taken off with a thigh injury, the match was abandoned at half-time. And rightly so: the livelihood and even the careers of professional footballers are at risk if they get injured.

Some sports are inherently dangerous, especially those labelled 'extreme', such as kite-surfing. This involves harnessing two powerful natural elements, the wind and the ocean. Competitors, using lightweight boards and kites up to 23 metres long, are judged on their wave riding, the height of their jumps and their moves in transition between jumps. Yet trying to harness the wind to demonstrate the tricks of the sport can be hazardous. Strong gusts can drag surfers through the water and into the air, only to plunge them back down into the sea in uncontrolled crashes, their safety gear of little use in the turbulent conditions. At least two kite-surfers have died in such accidents. Although less extreme, winter sports, reliant as they are on freezing conditions, remain an obvious area of injuries due to slips and falls. Statistics for the late 1980s actually suggested that skiing in Scotland was more dangerous than in either Austria or Switzerland. This was due to the ignorance of casual skiers and inadequate provision of piste markers and hazard signs by the ski-lift operators. Fears of litigation have improved signage but fractures, strains and sprains are still common among skiers and snowboarders.

Formula One fans love wet races in which the mobile advertisements that masquerade as cars shimmy and aquaplane around and across the track, their tyres at the limit of adhesion. Most drivers do not share this enthusiasm. The sport is dangerous enough without aiming a car into a wall of spray at over 300 kph. Too many have shunted the car in front, totally blind to its existence. In torrential rain a few drivers – courageous or crazy – thrive. The rest are content simply to survive. Yet collectively they have never refused to race. That decision has always been left to the chief marshal.

Not Playing the Game

Sports organisers can always take the decision to call off a sporting event, though sometimes they appear to be more conscious of the economic than the human costs. More generally, officials – referees in the football codes, umpires in cricket and stewards at horse-racing – have the authority to postpone or abandon a match or race if conditions are considered too dangerous. One unusual instance of this occurred at the start of the 1981 first-class cricket season when umpires suspended play at Fenners in the game between Cambridge University and Essex because of the cold. The chill from a biting wind prevented bowlers gripping the ball and one of the Cambridge batsmen, who wore contact lenses, could not see properly as his eyes were streaming. Umpires have also taken players off when lightning was around – as they did in a match between Northamptonshire and Essex in 1975 – but normally the darkening skies and accompanying rain are cause enough.

Sometimes participants, worried about safety, take matters into their own hands. Jockeys refused to ride at Beverley in 1989 and Haydock in 1996 when they felt the ground was too soft. In rugby, a Bath v Newport Heineken Cup match in 2001 was postponed just before kick-off as both teams refused to play. Bath had been advised that there would be no overnight frost and had left their Recreation Ground pitch uncovered. As it was, the temperature fell to minus 7°C and a hard frost set in – so much for the local weather forecast! In a Premiership game a local referee would have been called in to make an early morning inspection, but there were no clear rulings on this for Heineken matches. There was no inspection till Didier Mene, the French referee appointed to the game, arrived at midday. He felt that the pitch was playable but the two captains, Ben Clarke and Gary Teichman, voiced their opposition, believing that one end of the ground was too dangerous.

Volunteers from the capacity crowd held a tent aloft so that heaters could blast the hardened turf. Mene held a second inspection, again advocating that the match go ahead, but the players overruled him and the game was delayed for 24 hours. During a cricket match in July 1974 between students from Birmingham and Manchester universities, there was a blinding flash, a deafening bang and scorched turf where a lightning bolt struck the pitch. All but two of the players on the field were knocked over and, although no one was injured, there was a distinct reluctance to continue the game: it was abandoned.

Lightning is dangerous. In a thunderstorm, flashes travel at 144,000 kilometres a second carrying an electrical current of between 10,000 and 50,000 amps. When they hit the ground the charge can be carried several metres through the earth causing injury to anyone around. Those struck surprisingly rarely suffer significant burns as the electricity hits for only a millisecond or less. The major cause of injury is temporary cardiac or respiratory arrest. Several budding British tournament players counted their blessings as well as their scores in the fourth round of the PGA European Tour Qualifying School on 25 November 1997 at Guadalmina near Marbella in southern Spain. When a thunderstorm broke they ran for cover, but lightning struck the umbrella of Italian Alberto Biraghi and knocked him flat when he was only a few yards from shelter. Rushed to hospital after being hit, he could not move his legs for 20 minutes and still had cramp three hours later. The players behind him said he was alight from head to toe. The danger of lightning has been acknowledged by the Royal & Ancient Golf Club, the international rule-making body, who have decreed that competitors can stop play of their own accord if they feel conditions are too dangerous. This legislation followed the deaths of golfer Peter Kelly and caddie James Miles, who were killed when sheltering under an umbrella while playing in a torrential storm at Maidenhead in 1952. During the French Open at Le Golf National near Versailles in May 2000, Miguel Angel Martin refused to drive off the 17th tee because of the threat of lightning. John Paramour, the tournament referee, pointed out that the lightning detection equipment had not picked anything up. Martin acknowledged that he had not seen any flashes but was adamant that he had heard thunder and that was enough for him.

Those that participate in sport do so aware of the risks involved and, generally, with an implicit willingness to accept them. Nevertheless, concerns by officials about player safety, allied with fears of litigation for not observing their duty of care to participants, have led to sports authorities and organisers prescribing what should be done and proscribing what should not. In the mountains and hills, climbers and walkers are generally advised of dangers but left to make their own decisions. On the ski slopes, however, there are tight regulations. Many ski centres in Europe and the USA now ban people from carrying babies in backpacks after cases where the child has frozen to death, or had to have feet amputated after frostbite set in. At many ski resorts the proprietors have also gone to great lengths to try to ensure the safety of their customers with rigorously trained ski patrollers, carefully delineated boundaries, hot lines to emergency services, procedures to clear the hill when the weather closes in and a huge amount of expensive snow-clearing equipment. Yet, as in the Cairngorms, not all off-piste areas can be patrolled and so skiers venture there at their own risk.

The sea can be as dangerous as the slopes. Most regattas now utilise safety boats to render assistance to sailors in trouble, but this is not always possible, especially for offshore racing where vessels can be out of sight of each other for hours, sometimes days. Many organisers now have minimum safety requirements that must be adhered to, the level of safety equipment being related to the type of boat and location of the event. At the ocean-going level, safety gear is now almost as important as the vessel itself. The outcry following the tragedy of the 1979 Fastnet race in which helpless crews tried to survive hurricane conditions led to rules forcing all craft in long-distance races to carry a full range of safety equipment. But, of course, no amount of equipment can make up for human exhaustion.

9

Now is the Winter of our Discontent: 1963

I t was the year they dug carrots from the frozen ground with pneumatic drills. The year when the sea froze at Torquay in Devon and cross-channel ferries cracked their way through ice floes en route to Calais. The year when beer bottles burst with the cold and shoppers carrying bread rolls were attacked by flocks of starving pigeons. It was the winter of 1963.

The twentieth century witnessed few cold spells to match it. Beginning with the worst bout of freezing fog for a decade in December 1962, the weather turned wintry on the 22nd and continued in a similar vein until early March. Snow, frost, ice, gales, fog and bitter cold wreaked havoc on the sports programme and caused at least one sports writer to question whether 'the winter game' was an apt description of football in Britain. Not in 1963, it wasn't! By the first Saturday in February, 385 first-class matches and cup ties had been called off. There was very little rugby or hockey played until the beginning of March, when horse-racing also emerged from a ten-week lay-off. Meanwhile, cross-country runners found themselves making headlines in shrunken sports sections of newspapers desperate for news. The story of this sporting shut-down is one of ingenuity and ineptitude, frustration and fluke, delay and delight (although most of that was reserved for followers of minority pastimes such as outdoor curling in Keswick, skating on the Severn and skiing in suburban London). For most sports fans and competitors, it was a winter of discontent that lasted for nearly 12 weeks.

Week 1: 22–28 December
When fog wrecked the football fixture list on the last Saturday before Christmas, few would have believed that the next complete league card would not take place till 16 March. The murky weather resulted in 18 match postponements and a further 8 abandonments. On Christmas Eve snow fell in the west of Scotland and the final of the Glasgow Cup at Hampden between Celtic and Third Lanark was postponed. Further south, heavy frost put paid to the rugby league match between

Batley and Dewsbury, and a hammer and chisel had to be found to insert the stumps for the Christmas Day cricket match at Brighton between an Electricity Board XI and 'the Yule Logs'. Up and down the country, swimmers braved freezing temperatures for their annual Christmas Day races, although ice prevented the morning swim at Southport's open-air sea bathing lake and only a brisk wind kept the Serpentine from freezing over for the 99th annual Peter Pan Cup. Then the snow spread south. By Boxing Day afternoon it had reached London; by the 28th, there was up to 45 cm over much of Surrey and the South-east. The big freeze had begun.

Week 2: 29 December–4 January

The final Saturday of 1962 saw a limited football programme, particularly south of the border. Out of 45 scheduled English League matches, 35 were called off because of snow-covered or frozen pitches, but Scotland fared better with only seven games postponed, all in the lower divisions. Unfortunately, that was as good as it got for the next two months. Further snow and gale-force easterly winds hit the whole of England, Wales and southern Scotland on the night of the 29th and continued for up to 24 hours. The country awoke on New Year's Eve to snowdrifts, severe disruption to transport and little hope for New Year's Day fixtures. On 1 January, there was no English football, no racing and no rugby, although once again Scotland came off better, with a number of soccer matches going ahead over the holiday period. The prospects for the week's racing were bleak. Plumpton and Windsor were under 30 cm of snow and Leicester had 45 cm, while Manchester had escaped the blizzard, but not the frost.

One die-hard group of sportswomen tried to carry on. The festive season traditionally saw a number of regional hockey tournaments and this year was no exception. The Scottish Women's Hockey Association had got the ball rolling before the New Year with their annual inter-district event, but the snow-covered pitch at Perth made ball control and stick-work difficult. At the East Anglia tournament, the ball sometimes refused to roll more than a metre before transforming itself into a snowball. At Worthing, it failed to roll at all as the Southern Counties meeting was cancelled for the first time in over 30 years. The Western Counties hockey women considered playing on the beach at Weston-super-Mare as the pitches were covered in 45 cm snowdrifts. The Midlands Counties' teams decamped to Ramsgate, where their event finally began on 2 January, two days late and in Arctic conditions. The following day, with the pitches resembling skating rinks, it was called off again, but a slight thaw eventually allowed play to continue uninterrupted for two full days, albeit on a treacherous surface.

Football was less fortunate. By 4 January, 93 English League matches had been postponed or abandoned, and the story from grounds across the country was a gloomy one. In London, Arsenal's Highbury pitch was covered in 22 cm of snow and groundsmen were reluctant to clear it in case exposure to the bitter air made matters worse. There were 5 cm of snow at Barnsley, 10 cm at Norwich and 45 cm at Bristol. Crystal Palace and West Ham had given their players a day off and Queens Park Rangers were training in a local gym, but Burnley were hoping to have a full-scale practice on their all-weather pitch ahead of a third-round FA Cup tie with Spurs. Their efforts were wasted. White Hart Lane was under 35 cm of snow because the previous week's blizzard had arrived before the unwieldy plastic protective sheeting could be laid on the pitch. In Scotland there were mutterings in the press about the benefits of undersoil heating which kept Murrayfield frost free but had palpably failed to work at football grounds. (The installation of an electric blanket at Everton's Goodison Park in 1958 had led to considerable drainage problems, and other clubs were reluctant to invest in costly heating systems.) The Rangers manager said that electrical engineers were keeping them informed of new developments. The chairman of Hearts, unable to play their last four games, was more concerned about immediate solvency, describing the present situation as 'financially disastrous', a phrase which would be much used in the weeks to come.

Week 3: 5–11 January

The first Saturday of the New Year normally sees an exciting third round of the FA Cup – but not this year. Only 3 out of 32 matches went ahead, and by the end of the week the figure had crept up to 6. They were played on extraordinary surfaces, ranging from heavily sanded rolled snow to slippery soft mud with pools of water at the corner flags. Alf Ramsey, the manager of Ipswich, saw his team beat Mansfield on a pitch ankle deep in snow and called the conditions 'farcical'. There were, however, a few sporting oases of light in a desert of despair. The cross-country running season got off to a reasonable start with only 3 out of 15 county championships cancelled. On a bitterly cold day, snowploughs and tractors cut paths for the races at Middlesex, Somerset and Essex, and competitors wheezed and skidded their way round, collecting cuts, bruises and sprained ankles for their pains. After two inspections, the National Hunt meeting at Ayr also went ahead in spite of fears that trainers from the north of England would be unable to travel. Neville Crump, based in Yorkshire, found that his route across the border was still blocked by snow and scratched his 12 runners from the meeting. But, although many horses had not been beyond the stable yard for two weeks, 60 turned up for the six-race card. The winner of the final race was Cinzano – but that was the last punters had to cheer about for the next eight weeks as racehorses remained idle for an unprecedented period.

With the arrival of hard frosts after the weekend, mainstream sport virtually ground to a halt. There was no rugby, no racing and precious little football. Several snow-clearing machines were tried out at football grounds with varying degrees of success. At Derby County attempts were made to thaw the pitch using a gas-heated contraption which blew hot air through a funnel. In Dundee, both football grounds were cleared using an American snow blower but the pitches underneath were found to be covered in thick ice. Ice on the River Cam prevented the Cambridge rowing eights from covering more than one kilometre in open water.

Fans of winter sports came into their own. Outdoor curling was played at several venues in Scotland and the north of England, and there was talk of the Grand Match, the world's major outdoor tournament, taking place in Fife. Meanwhile the temperature at Grantown-on-Spey plummeted to minus 21°C on 7 January, bottles of lemonade burst with the cold and diesel oil froze in several buses. Sheep buried in snowdrifts on Dartmoor were being eaten by foxes which burrowed through snow to get at them. To cap this silly season, *The Times* resorted to telling the heart-warming tale of a duck, found by workmen at a brick works in Bedfordshire, frozen solid and presumed dead. They thawed it out, and gave it a hot drink; apparently it was soon quacking loudly!

THE SAGA OF SAN ISIDRO

When 25 players and 5 officials from the Club Athletico de San Isidro, rugby union club champions of Argentina, arrived in Britain on 10 January their eight-match tour had been carefully planned. The first Argentinian team to play in this country was largely made up of law students who had paid their own expenses of almost £600 (£7,500 today) per player to take on sides such as Bridgend, Richmond, Old Paulines and Bedford. They had arranged two matches in Scotland and one in Belfast, with the tour kicking off at Wasps on Tuesday, 15 January. It didn't quite happen like that! On the appointed day, they found themselves 500 kilometres from the icy fields of north London, preparing to meet a team from Penzance in Cornwall, the only part of the country which was not completely snowbound. In front of a large and enthusiastic crowd, they won 12–6. Unfortunately, from then on it was all downhill.

The next two weeks saw them kicking nothing but their heels as match after match succumbed to the weather. First the game at Richmond was cancelled, then Bridgend (but they were taken to watch Wales v England at Cardiff Arms Park instead), then Old Paulines (but they were entertained to a 'turkey dinner dance'), then Kelso (but there was a civic reception in their honour). They were learning a lot about social niceties in Britain but not much about rugby! Ebbw Vale magnanimously offered the Argentinians a match under floodlights whenever the weather permitted, with the tourists guaranteed the entire gate money to help cover their expenses. It never happened.

Finally, on a bitterly cold 29 January, they took the field at Bedford and lost 6–0 on a pitch only partially cleared of snow. Next stop Murrayfield, the only playable venue in Scotland thanks to its electric blanket, which had just staged the Five Nations international between Scotland and Wales. There, short of training and perhaps slightly overawed by the surroundings, they went down 21–9 to Edinburgh University. They had been in Britain for over three weeks and had played only three of the seven matches scheduled to date.

At this point Ireland came to the rescue, offering games with University College, Dublin and Tullamore, but poor San Isidro found that they had swapped the frozen fields of the mainland for glutinous Irish mud. Bogged down on a heavy pitch, they lost 26–3 to the University but then managed two creditable draws, the second against Queen's (one of the few pre-planned games which actually went ahead, although it had to be played at Carrickfergus because the Belfast pitch was under snow). Their farewell to Britain took place on 16 February in a hastily arranged fixture against London Irish. They lost 17–15 . . . but the pitch was nearly perfect!

Week 4: 12–18 January

The bizarre events of the football season continued unabated. Saturday saw 41 of 46 English League matches and 29 of 32 first-round Amateur Cup ties postponed. The game between Windsor & Eton and Eastbourne went ahead after players and officials, together with their wives and girlfriends, had cleared the pitch. The Eastbourne team took the field wearing tracksuit tops and gloves, and were noisily directed from the trainer's bench by a 'well-wrapped gentleman in climbing boots'. They won 3–0. Professional clubs were urged to play outstanding FA Cup ties on weekday afternoons instead of evenings, partly to avoid the lower temperatures and partly to avoid power cuts which threatened the use of floodlights. The FA offered interest-free loans to those clubs in dire financial need, while the wealthier contemplated taking their players abroad. Although the Isle of Man seems unlikely to have been uppermost in their thoughts, the Manx parliament offered over 20 snow- and frost-free football grounds to British teams for training purposes. Swindon were kitted out in yellow gloves and gym shoes for their 5–0 defeat of QPR, whose manager complained that the pitch was not good enough for professional football 'as entertainment'. Spurs finally played Burnley in their postponed cup tie and lost 3–0. The pitch was treacherous after a light snowfall on top of frozen ground, and the conditions enabled Burnley to score two freaky goals, rendering one of the more mouth-watering ties of the round nothing more than a lottery.

Meanwhile, outlandish tales continued to fill the newspapers – of sailing on ice on the Norfolk Broads, skiing on the frozen Ruislip lido and skating on the River Severn at Shrewsbury. Tobogganing took place on golf courses across the south of England. Photographs showed the sea at Herne Bay in Kent frozen 100 metres out from the shore, and racehorses from Epsom exercising on an ice-strewn beach in Sussex. The annual amateur golf competition for the President's Putter went ahead, not at its usual venue of Rye, where the course was covered in snow, but along the coast at Littlestone. There were rumours that the clubhouse had run out of port by 3 p.m. on the first afternoon and that players, wearing five sweaters to fend off the cold, were being welcomed at the ninth green by the Golf Society's captain, bearing brandy like an alpine St Bernard. 'Business as usual' was exemplified by those guardians of true grit, the Rugby Football Union (RFU) and the Cambridge rowing crews. The rugby officials, notwithstanding wholesale match cancellations by the clubs, insisted on holding a final trial at Torquay between England and The Rest on an icebound and partially straw-covered pitch which mocked any attempt at skilful play, and could have told the selectors nothing except whose studs had the best grip. The university rowers, frozen out of the Ouse at Cambridge and Ely, went north to Earith in Huntingdonshire where the river was just about tidal and the movement of the water helped to break up the ice. Rowing remained perilous, with blades regularly descending on ice floes and banks of fog hampering visibility.

There had been a slight thaw in Scotland, which encouraged hopes that the salmon rod-fishing season might open as usual on the River Tay but the curlers still managed to have their day. Ice on the Lake of Menteith held firm at nearly 15 cm, allowing the sixth Grand Match of the twentieth century to take place between the curling armies of the north and the south. While London registered its 25th consecutive day of 3°C or below, another twentieth-century record, 1,800 players travelled from all parts of Scotland over icy roads, in blizzards and through snowdrifts to join the festivities. With due ceremony, a cannon signalled the start and, in front of BBC television cameras, the curlers enjoyed three hours of match play in a scene resembling a winter landscape painted by the elder Brueghel. Stones rumbled over the lumpy ice while players were spurred on equally by the warmth of braziers on the lake shores and the liquid contents of many a hip flask. The north won by 2,560 shots to 2,538.

Week 5: 19–25 January

Unbelievably, matters got worse. Fresh blizzards arrived in time for the next weekend and most sport succumbed again. Comparisons began to be made with the winter of 1947, the snowiest of the century. In England, the total of postponed football matches rose to over 200: 176 league and 25 FA Cup ties, a situation unparalleled in the history of the game. When Spurs beat Blackpool 2–0 at White Hart Lane, players had to peer through driving snow to see colleagues, and keeping the ball stationary at set pieces was nearly impossible in the gale-force winds. Charlton v Southampton was abandoned ten minutes into the second half after the lines became covered in snow. According to reports, the match should never have gone ahead. The pitch, though heavily sanded, was so slippery that both goalkeepers fell over whenever they took goalkicks, and it was fortunate that no one was seriously hurt. Chelsea did the sensible thing and left the country, a squad of players departing for Malta to play a couple of friendlies there.

Ground-clearing experiments continued throughout the week, with limited success. The latest effort at Dundee's Dens Park involved the use of plastic sheeting anchored to the ground, while hot air was forced underneath. The surface snow and ice melted but, with the subsoil frozen hard, the water was unable to drain away and ice soon re-formed. A similar trial at Leicester's Filbert Street also ended in failure, with the inventor of the equipment protesting that his polythene tent and hot-air equipment was designed for prevention, not cure. With 35 cm of frost in the ground, he complained, the test was unfair. More conventional methods worked better. A machine which broke up the packed snow and ice, enabling it to be swept and shovelled, was used at Watford and at Harlequins' Teddington rugby pitch. Good old-fashioned braziers, stoked throughout the night by club officials, ensured that the Scottish Cup first round tie between Gala Fairydean and Keith went ahead at Galashiels. Gala won 4–0.

There were mixed fortunes elsewhere. The rugby international between Wales and England was played at Cardiff Arms Park but proved to be a considerable ordeal, both for the players on a skating rink of a surface and the 46,000 crowd on the bitterly cold terraces. The Oxford rowing crews, practising at Henley for the university Boat Race, came ashore with their tracksuits and oars caked in ice and the swivels on their boats frozen solid, 'like British Rail points'. The spray thrown up by the bows of their accompanying launch froze into needles of ice. The rowing correspondent of *The Times* reckoned that extreme cold and discomfort had much the same effect as exhaustion and, since the Boat Race was said to be character forming, the recent ordeals should have taught the crews valuable lessons, as well as helping coaches with selection problems. Cambridge, meanwhile, having given in to the pack ice at Earith, began training on the River Nene at Peterborough, where hot water discharged from a nearby power station kept the water ice free. On the roads, 10 out of 65 British entries for the Monte Carlo Rally were knocked out by blizzards en route to Dover from the starting point in Glasgow. Those who boarded the *Maid of Kent* ferry heard from her captain, as the ship battled through ice floes, that he had never encountered such conditions in 30 years of Channel crossings. This was also the week in which newspapers carried photographs of workers harvesting carrots in Suffolk with the aid of pneumatic drills. Some were damaged in the process and had to be broken from clods with a hammer, but the shortage of root vegetables still made the operation worthwhile. The misery continued and the forecasts were not encouraging.

Week 6: 26 January–1 February

History was made on the last Saturday of the month, when, after weeks without any Pools coupons, the first ever Pools Panel of experts forecast the week's football results. (An ongoing legacy of winter 1963, we have devoted an entire chapter to this ground-breaking, weather-induced phenomenon.) A few games went ahead, including nine Scottish Cup ties, but playing conditions were ridiculous. When Brechin City lost 2–0 at home to Hibernian, the visitors were greeted on arrival at the ground by the sight of 20 volunteers trying to spread several 60 cm-high dumps of sand across the pitch to mop up the pools of water. Others busied themselves constructing little walls of sand around the touch-lines to prevent cleared water from seeping back. The match report suggests that players spent as much time prone in the mud as attacking the opponents' goal. South of the border, Swindon beat Luton 2–0 at Kenilworth Road in front of a small and subdued crowd, on a pitch coated in thick orange sand. Turning was hazardous and tackles could only be attempted with care. Portsmouth drew 1–1 with Scunthorpe in a third-round cup tie, the last result anyone wanted in the circumstances, on a pitch dressed with peat. Players were unable to control any forward momentum on the icy surface and frequently crashed heavily into each other. Many clubs simply gave up trying to produce usable pitches and decided to await improvements in the weather. Some waited . . . and waited . . . and waited, as did the FA, postponing the fifth round draw for a week. Chelsea, however, informed that Stamford Bridge was still unplayable, wisely decided to remain in Malta and arranged another match there. Others did likewise, with Manchester United and Coventry organising a friendly in Dublin.

On the positive side, sports fans were no doubt delighted to hear that Cambridge University had become the first club to stage a lacrosse match since Boxing Day, and that members of the RAF spent an hour defrosting a plane in Oxfordshire to enable their rugby players to fly to Cornwall, one of the few frost-free counties in England, for some match practice against local side, Penryn. The Cold War began to take on a new meaning!

Week 7: 2–8 February

Six weeks into the long freeze the consequences of the severe winter for British sport were beginning to show. For many football clubs the worst problems centred on increased costs coupled with reduced income; for bodies such as the Football League, it was the logistical nightmare of rescheduling matches. Clubs were caught between the need to generate income through the turnstiles and the dangers, both physical and financial, of playing on barely fit surfaces. A lucrative cup run could be jeopardised by a pitch on which skill played little part and inferior opponents got lucky. A star player might be injured. The number of paying fans might be fewer than average, given the miserable conditions for spectators and the inconsistent quality of soccer on show. When the extra costs in manpower, sand and machinery to render the pitch playable might be rewarded with a poor crowd, a defeat or both, it is no wonder that some clubs reduced their efforts to beat the winter. In addition to the pitch, their responsibilities extended to icy or snowbound terraces, frozen toilets and side streets where heaps of snow made parking impossible. Sometimes the best option seemed to be to wait for warmer weather, which was bound to arrive . . . eventually!

For football administrators the backlog in matches had reached massive proportions. With 286 unplayed games on 2 February – the equivalent of five and a half full Saturday programmes – there was no option but to extend the English League season by two weeks, to 18 May. The United Counties League was granted permission to play Sunday football to cope with its fixture pile up. The backlog of FA Cup ties, standing at 99, ensured that the fifth-round draw was postponed again. For some leagues, however, extensions were impossible. The Southern Amateur Football League, consisting of 34 clubs in three divisions, had to abandon its season because there was no chance of completing fixtures by the closing date. Most clubs were associated with cricket teams, and therefore an extension into the summer was not an option.

For rugby league the hardest winter for 15 years could not have come at a worse time. The sport's ruling body, concerned at the decline in attendances, had embarked on an experiment in season 1962–63, dividing the league clubs into two divisions. More meaningful and competitive fixtures were expected to increase spectator numbers and improve finances. But hardly any matches were played from January to March and by then nearly 200 had been postponed. There was no alternative to extending the season. Elsewhere, a committee of the RFU chose not to follow suit, a decision made easier by the absence of leagues at this time, although there were protests from clubs such as Northampton, which had lost £1,000 (£12,500 in current prices) in gate money to date. The Badminton Horse Trials, scheduled for three days in mid-April, were cancelled in February because no preparatory work could be carried out on the cross-country courses. Organisers hoped to substitute a one-day event instead. Another casualty was Scottish District League hockey. After ten blank weeks, most competitions had been abandoned by the first weekend in March.

Amongst other sports, horse-racing was hit particularly hard: 63 meetings had been lost so far and there was no end in sight. Although courses affected by the bad weather had all been granted an extra fixture, this was scant compensation, particularly since few had bothered to take out insurance against abandonment. For owners, it was a frustrating time, paying bills for animals confined to stables, chomping their way through sacks full of expensive fodder with no possibility of earning their keep. For trainers, exercising horses was an endless headache, especially when snow gave way to heavy frost and iron-hard ground. For jockeys, it was brutally simple – no work, no pay.

Week 8: 9–15 February

As the year grew older and the days grew longer, the weather became more fickle. For the eighth successive Saturday it destroyed the football league programme, with a carpet of snow and ice in the north and thawing, flooded grounds in the south. Seven English League fixtures survived but the entire Scottish programme was wiped out. The Pools Panel met for a third time, parts of Scotland suffered their worst snowfalls of the winter and men's hockey finally made an appearance in the sporting calendar, thanks to an all-weather pitch at Devonport, where the Royal Navy beat Western Counties 1–0. The only winner in the rugby international at Lansdowne Road was the weather as the Irish and English packs ground out a scoreless draw, the first since 1910. After days of rain, the pitch was described as a 'glutinous morass' and long before the end of the match it was impossible to distinguish between the players, who were all caked in mud.

In order to speed up the progress – a misnomer if ever there was one – of the FA Cup, clubs were encouraged to play ties at any available neutral venue, a suggestion which met with no enthusiasm whatever. The fifth-round draw was postponed for the third time, with 48 names scheduled to go into the bag for 16 places. In the Amateur Cup, 53 clubs vied for 16 slots and no date was planned for the draw. In Scotland, by a coincidence which had nothing to do with the atrocious weather, the 37 Scottish League clubs were asked to consider a proposal for summer football, starting in June with a winter break from the beginning of December to the first Saturday in March. The vote at the end of the month was keenly awaited, given the problems of the current season.

Week 9: 16–22 February

The largest football league programme of the year got under way on Saturday, 16 February – a total of 11 matches! Bolton played for the first time since 8 December, the longest period in their league history without a game, and lost to Arsenal at Highbury. But Charlton v Cardiff, postponed ten times to date, was off again, the slow thaw having caused too much slush on the pitch. Cardiff had already travelled needlessly to London twice, incurring rail fares of £300 (£3,750). (It was third time unlucky on the following Monday when, at the 12th attempt, they lost 1–0 on a bitterly cold night.) The FA Cup final was rescheduled for 25 May, three weeks late, and the English League season was extended again. Somewhat churlishly, given the rigours of the winter, the secretary of the MCC said that this was very unfortunate for cricket and hoped that it would not set a precedent. With the continued fixture congestion and concern about international commitments in late spring, several representative matches were cancelled, including the Scotland v England Inter-League game. Apart from wartime, this had been staged every year since 1892.

Rugby was one sport that took on a more hopeful aspect, with several games played in the London area, although only after volunteers had dug and swept pitches clear of ice and snow. The annual Hospitals' Cup rugby tournament finally began on the date originally set aside for the semi-finals. Cross-country athletes continued to compete over sheets of ice and through pools of water, but many sports remained frozen out. Anglers made holes in the ice on reservoirs in which many fish had died through lack of oxygen. The first pre-Olympic training event arranged by the British Hockey Board at Lilleshall was postponed till April because of adverse ground conditions. Fears began to be expressed for the National Hunt Festival at Cheltenham, less than three weeks away. Snow had been largely cleared, by hand as machinery would have damaged the turf, but the ground was still frozen and night frosts reversed any daytime thaws. Wetherby racecourse boasted 15 cm of snow with deep frost and Chepstow lay under massive snowdrifts. One consolation was that the anticipated rise in temperature posed no risk of flooding. A lack of rain, particularly in western parts of the country from Fort William to Cornwall, meant that rivers were running at or below summer levels and water tables were low. Fishing was at a standstill on the Rivers Dee and Don, and nearly all salmon catches elsewhere were in the lower reaches. As it was too shallow for fish to get upstream, salmon were being netted at the mouths of rivers and good catches were being made on the Tay within the city of Perth.

Week 10: 23 February–1 March

For the first time in five weeks the Pools Panel was not required to sit, but many games took place in deplorable conditions. Spurs beat Arsenal at Highbury on a snowbound pitch and Fulham beat Nottingham Forest on a surface which quickly became heavy and very slippery. By Monday evening, severe frosts had led to further postponements. The FA Cup third-round tie between Birmingham and Bury was called off at half-time when ice began forming on the touch-lines, and the referee abandoned the replay between Third Lanark and East Fife at full-time with the scores level (1–1) because it would have been too dangerous to play extra time. Grounds such as Ibrox, which ran east to west and were shielded from the sun by a grandstand, displayed pitches half playable and half covered in thick ice. At Dens Park, the latest attempt to clear the ground was thwarted when an icebreaker, having managed to clear the sunny side, broke two of its teeth on the icy hard area below the stand! This was particularly unfortunate as Dundee, without a First Division game since 5 January, were due to play a European Cup tie against Anderlecht in Brussels on 6 March and were desperate for match practice. In spite of all the winter misery, the Scottish League clubs voted to retain the existing football season, dividing largely on east/west lines (no surprise there!) with Glasgow clubs against a change and those from Edinburgh and Dundee areas in favour.

Rugby internationals at Twickenham and Murrayfield went ahead in bright sunshine, but at Northampton a game between East Midlands and the Barbarians was called off because there were still 12 cm of snow on the pitch. The first falls had occurred ten weeks previously. There was still no racing or hunting. Point-to-point seasons were postponed, often because country roads were still snowbound, and farmers were said to be exercising steeplechasers by using them to round up sheep. Spring was not in the air.

Week 11: 2–8 March

The first week in March produced a bag of mixed fortunes. At the weekend, the total of lost race meetings topped 100 and the National Hunt Festival remained in jeopardy because of continued night frosts. Plans were made for the Champion Hurdle and the Gold Cup to be staged in early April. By midweek, however, the long-expected thaw, together with a little rain, gave renewed hope for Cheltenham and the possibility of a meeting at Newbury on Friday and Saturday. To the delight of the racing world, but particularly the owners and trainers of hundreds of fat horses, a seven-race card was held, with several fields of over 20 jumpers. A full rugby programme took place in England, although Leicester, without a game since 15 December, had to switch their match with Harlequins from a rock-hard Welford Road to the more suitable conditions at Teddington. They still won 15–8. A record field took part in the England cross-country championships at Cambridge on a course covered in slush, ice and water. Skiing conditions in Scotland were very good, but lack of rain and extreme frosts had left tracts of moorland tinder dry and susceptible to fire. At St Andrews, three of the golf courses opened for the first time on a Sunday. Over 400 golfers played, two-thirds from Glasgow, Edinburgh and the Borders, where many courses were still unusable.

Finally, after 14 postponements and 1 abandoned match, Birmingham and Bury managed to complete their third-round cup tie; after a 3–3 draw, Bury went on to win the replay. Not a blade of grass was visible when Leyton Orient beat Derby 3–0 on a sea of mud, and it oozed nearly 3 cm deep at Upton Park when West Ham overcame Swansea in their fourth-round tie. By Wednesday, postponed matches were being played at the rate of six a day; the fifth round of the Cup was finally drawn – 36 names for 8 ties – and Charlton took on Chelsea on a balmy evening in front of a crowd of 37,000. Chelsea won 3–0. The football nightmare appeared to be ending.

But the thaw did not herald good news for everyone. More than 800 staff at Littlewoods' Pools on Merseyside lost their jobs, with the weather cited as a major factor in the branch closure, while the south of Scotland and the Borders experienced their worst floods for seven years.

Week 12: 9–15 March

On a Saturday of wind and rain rather than snow and ice, the football league programme approached normality, only Watford v Barnsley having to be abandoned at half-time. Racing at Newbury continued, but gale-force winds disrupted Boat Race training on the Thames, and two dozen crews practising for the Head of the River race became waterlogged and sank. The final third-round FA Cup tie was played to a finish on 11 March, over nine weeks late, when Middlesbrough beat Blackburn 3–1, and the Scottish Cup second round was completed on 13 March with Partick Thistle beating Arbroath 3–2 in a second replay. The Scottish football season was extended to 18 May and midweek games were scheduled for Monday as well as Wednesday evenings to cope with the backlog. Some teams found they had four fixtures in the space of eight days. The Hospitals' Cup rugby tournament, starting six weeks late, reached its climax. The National Hunt Festival went ahead, although conditions were not ideal. There were hazardous greasy strips on the take-off and landing sides of the fences, months of snow having killed off much of the grass, and car parks were awash with mud.

Postscript

Fears that the winter would end dramatically, as it had in 1947, with gales, heavy downpours and the worst ever river floods, proved groundless. Although a few areas where snow still lay thick suffered flooding, over much of the country winter quietly faded away. But the spring was a wet one, and the legacy of winter continued to make life difficult for sports that required a reasonably dry playing surface. Golfers taking part in the Schweppes Championship at Royal Birkdale in early April had to deal with pale greens, bare in places, on which the grass seemed to have died. Four putts were common, one unfortunate competitor took five, and the winning score of 306 was reckoned to be the highest of any professional tournament since 1919. A Badminton one-day event was held in place of the usual three-day horse trials, but the going was treacherous even before driving rain on the day, and only half of the entrants finished the cross-country course. In rugby union, a few clubs tried to make up some of the lost fixtures by playing midweek and evening games but most just gave in to the inevitable. The wet April weather didn't help. When Bristol RFC staged a 75th anniversary match against a 15 selected from the forthcoming tour party to the Antipodes, no grass was visible between the 22-metre lines, 'only an area of viscid mud'.

Eventually most sport won through. In football, the pre-freeze leaders, Everton, went back to the top of Division One and clinched the title on 11 May, two weeks before the rescheduled end of season. Leicester, overcoming weeks of inactivity and some dubious hot-air machinery at icebound Filbert Street, made it to the final of the FA Cup where they were beaten 3–1 by Manchester United. In Scotland, Rangers did the double without the necessity of undersoil heating and Dundee, plagued with problems at their Dens Park ground, triumphed over adversity, reaching the semi-finals of the European Cup before going down 5–2 on aggregate to Milan. Trainer Fulke Walwyn surmounted all the winter problems by sending out Mill House to trounce 11 opponents by 10 lengths in the Cheltenham Gold Cup, but attendances for the National Hunt Festival were down on previous years because of uncertainties about the weather. The rugby league season staggered on till 1 June when Swinton were finally crowned First Division champions, but the experimental restructuring was abandoned during 1964. And, despite traipsing round half of East Anglia in search of clear water, Cambridge still lost the Boat Race to Oxford.

10

Dust to Dust: The Drought of 1976

It crept up on us gradually, stealthily, over more than a year. Its impact depended very much on where you lived. Its long-term effects, at least for the world of sport, were limited, but no one who saw the media coverage at the time would have thought so. It dominated the television news bulletins and the nation's conversation during a long, hot summer – and then disappeared overnight.

The great drought of summer 1976 had been building since spring 1975. Water shortages were experienced in south Wales, parts of the West Country and south Yorkshire as early as June that year, which turned out to be the fifth driest of the century. Standpipes were erected in the streets of Sheffield during the autumn, and householders were only spared when rain fell in November and December. Other parts of the country saw less than half their average winter rainfall. By spring 1976, water authorities gravely pronounced that the last fortnight in April was the final opportunity to replenish the water table – after that, most rain was absorbed by surface soil, which was already exceptionally dry.

THE WELL RUNS DRY

There are different definitions of drought. Meteorologically, drought is lack of rain (or snow). A dry spell is defined as 15 consecutive days, none of which has more than 1 mm of rainfall. A partial drought is 29 consecutive days when mean rainfall is 0.2 mm or less. Absolute drought is a period of 15 consecutive days with rainfall of no more than 0.2 mm in total. During the summer of 1976, parts of Devon suffered 44 days of absolute drought and at Edgbaston, home of Warwickshire County Cricket Club, there was no rain for 37 consecutive days.

A hydrological drought occurs when there is insufficient water supply to meet demand. In Britain, it usually refers to a winter rainfall shortage which fails to make good the soil moisture deficit of the previous summer or adequately replenish reservoirs.

The press descended into doomsday mode, conceding that there was little regional water authorities could do about the drought but querying their decisions in shrill tones. How come householders in the Anglian water region were being urged to conserve supplies when nearly 7,300 litres – the equivalent, apparently, of 80 baths – had been pumped into the water jump at the Woodland Pytchley Hunt point-to-point course, shrieked the *Sunday Times* on its front page. (The Water Authority retorted huffily that the Hunt was a commercial enterprise and paid for its water; therefore it was not in a position to deny access.) Elsewhere, racecourses were already being watered and there were wholesale withdrawals from the Windsor Horse Trials because of the hard ground. The problems for sport had begun.

In early May, the Middlesex rugby sevens were played in sweltering heat, the Benson & Hedges cricket cup ties in burning sunshine. Some anglers were already expressing concern about the forthcoming season. There were fears that falling water levels would kill off food supplies for many fish, particularly in stretches of shallow water such as canals, and that pollution would worsen as river flow was reduced. On the other hand, those who fished in large reservoirs were expecting a good year. With a drop of up to four metres in normal water levels, anglers could now walk across newly exposed areas to fish in virgin territory. As some lakes had shrunk by 20 per cent, fish stocks were now concentrated in smaller areas and a bumper catch was predicted.

By the end of May, statistics showed that the previous 12 months had been the driest on record. Four weeks later, the dry weather was accompanied by some of the highest temperatures ever recorded in Britain, reaching a peak of 36°C at Cheltenham on 3 July. What was particularly unusual was the number of days in which the thermometer rose above 30°C, including an unprecedented sequence of 15 days in late June and early July, which virtually coincided with Wimbledon fortnight. There were fast courts, worn grass and younger champions than in 1975. One commentator reckoned that all the older competitors, apart from the mixed doubles pairing of Françoise Durr and Tony Roche, had wilted in the heat. In London, roads melted, people fainted. The nation, fuelled by the usual press hyperbole, was in a state of rising hysteria. The tabloids collectively racked their brains to produce headline variations on 'Phew! What a scorcher!' Sunshades were erected for penguins at Dudley Zoo. At least 15 people were drowned in one weekend, prompting the Home Office to issue warnings about high tides, special care of small children and too much food and drink before swimming, which most people ignored as they headed for the coast or the nearest stretch of cool water. As the summer wore on, that became more difficult to find as reservoirs dried out and rivers dwindled to a quarter of their normal flow.

It wasn't the same everywhere. Although the Spey between Grantown and Cromdale was 350 mm below its normal summer level and a hosepipe ban was operating throughout Grampian region, Scotland in general was suffering far less than England. Reservoirs were well filled, some near to overflowing. Snowploughs had been seen in July (spraying sand on the melting tar of the A9 between Perth and Inverness!) and there had been large forest fires, particularly at Carrbridge near Aviemore, but most of Scotland had experienced steady rainfall in the spring and early summer, as cricketers were well aware. Rain had already ensured that, by 4 June, only two out of twelve cricket matches in the East of Scotland League had been completed.

The weather forecasters consistently got it wrong. The long-range forecast from mid-June to mid-July had predicted average sunshine and rainfall in the south of England, with cool, less settled weather and below average sun in Scotland. They had certainly not envisaged a heat wave, nor the complete absence of measurable rainfall in the London area between 20 June and 14 July. A senior member of the long-range forecast team at Bracknell defended his colleagues: 'We have only been going for 12–15 years and are still a young, pioneering science.' Unfortunately, the pioneers continued to miss the trail. Their prediction for August suggested unsettled conditions at first, with warmer, mostly dry weather later. It hardly rained anywhere in England for the entire four weeks. September was expected to be dry almost everywhere with rainfall below average, and temperatures and sunshine higher than normal. Guess what happened?

But more, much more, of that later. In the meantime, summer sport in Britain had to adjust to unexpectedly warm, sunny days. Golfers struggled in the heat. 'I'm exhausted,' said Tommy Horton, winner of the Uniroyal International golf tournament at Moor Park, Rickmansworth. 'My legs are gone.' So were jackets. It is said that, for the first time in the history of Lord's, MCC members were allowed to remove blazers in the pavilion – but ties still had to be worn! Thereafter, most cricket was played in hot summer sun prompting comments such as 'baking for the crowd and grilling for Glamorgan' as Nottinghamshire beat the home side at Swansea. The heat also resulted in record scores. Warwickshire made 287 in a Gillette Cup tie, the highest total to date for any side batting second in a one-day match. The third Test at Old Trafford was, of course, an exception. Rain inevitably stopped play there, though not in time to save England from an embarrassing defeat by the West Indies, who bowled them out for 71 in their first innings and won by 425 runs.

Early July saw the return of the Open golf championship to Royal Birkdale where there were record crowds and record temperatures; the English golfer, Peter Oosterhuis, joint seventh in 1975, was forced to cut short a practice round because of the heat. There was no shortage of water as greenkeepers were able to operate 40 sprinklers from a large storage pond, but some areas of the course still turned brown. Players found themselves having to contend with fast-running fairways and speedy greens more reminiscent of Augusta National. Well, the balls careered across the surface in a similar fashion but the comparison ended there. Instead of manicured verdant swards, Birkdale could only offer blotchy chequerboards of green and brown. According to the golf correspondent of a Scottish newspaper, Jack Nicklaus three-putted the first two holes because he was unable to fathom the change of pace as the ball sped from green to brown to green again. Tom Watson likened the effect to a roller coaster. Tom Weiskopf mused that Britain seemed to have been caught out by the heat. It was hard to sleep or keep cool, he said, because there was no air conditioning – and no ice! Perhaps surprisingly, after all the complaints, it was an American, Johnny Miller, who won.

Yachtsmen also had mixed fortunes. The hot weather often produced windless days and threatened to scupper competitions. A fleet of 116 Laser-class yachts was becalmed for two days at Torbay during the European championships in early July. After four days, there had been only two races and, since a minimum of four were required to constitute a championship, the event was fast running out of time. Fortunately, the wind picked up on day five to give perfect sailing conditions, which was more than could be said during Cowes week at the beginning of August. With a little less wind every day, racing there virtually petered out.

The government finally entered the fray, introducing an Emergency Powers Bill which would allow local authorities to regulate water supplies and phase in controls on non-essential uses of water. Industry and agriculture were to receive priority, while the general public was urged to cut down on the use of dishwashers and washing machines, and restrict baths to a depth of 12 cm. Sports grounds, racecourses and golf courses were at the back of the queue, but the Minister for Planning and Local Government, John Silkin, did not expect authorities to impose a total ban on them. It was not his intention, he said in the House of Commons, to ruin the wicket at Lord's or the Centre Court at Wimbledon.

Over the next few weeks, the Drought Bill was debated in Parliament, with all parties agreeing that measures had to be taken now to safeguard water supplies for the autumn. Under existing legislation, the government was unable to prevent the watering of council flowerbeds or private golf courses, to the obvious annoyance of householders, some of whom were already suffering hosepipe bans. The Central Council of Physical Education was alarmed about employment if all sporting activities were cut off. The Jockey Club expressed concern, and Lord Oaksey (later of Channel 4 Racing fame) put in a plea for the industry. There was limited sympathy. Clement Freud, a Liberal MP, asked the Minister to look into the abuse of water on racecourses such as Ascot and Newbury which gobbled up 'millions of gallons' each year. Dennis Howell, the Minister for Sport, rejecting a call for a blanket exemption from water restrictions for sports organisations, denied that the government was trying to put clubs out of business, but warned that commercial enterprises such as racecourses would have to make changes, including switching or postponing fixtures.

Horse-racing was undoubtedly suffering. Although good going could be found in areas where watering was possible, some racecourses had deteriorated from good to firm and from firm to hard. Trainers refused to run their most precious charges either for fear of damage on the course or because the animals could not be adequately prepared on the home gallops. Ryan Price, doyen of flat trainers, complained that he couldn't even canter juveniles at his Sussex stable. Some high-class youngsters hardly saw a racecourse during their two-year-old season. Fields, especially in National Hunt races, were often reduced to two or three horses, and flat race meetings at Brighton and Pontefract in early August attracted less than six runners per race on going labelled respectively firm and hard. With fewer horses, bookmakers also found the going hard. Shares in the Coral Leisure Group fell when the bookmaking arm of the business reported a drop in earnings compared with the same period in 1975. The heat wave and a shortage of runners were blamed.

High temperatures, too much sun and lack of rain soon began to exert a more negative effect on other summer sports. The English Bowling Association estimated that, by the middle of August, more than half of the 6,000-odd greens in England had been shut down, many burned out. Residents of the London borough of Southwark were lucky. The three municipal bowling greens there were being watered because it would have cost the local authority £9,000 (£39,000 today) to replace them if the grass died. Grass court tennis tournaments at Torquay and Malvern were cancelled. Golfers were playing in trainers in an effort to preserve fairways that were crumbling under spikes. Some courses were shut, others were discouraging visitors. Smoking was widely banned to prevent fire but the sunny, rain-free weather alone was quite capable of killing fairway grass, putting surfaces, rough and young trees without assistance from humans. Hundreds of courses, mostly in the south and west of England and south Wales, relied on the public mains for their water supply, and, as reservoirs evaporated, their future looked bleak.

Meanwhile, the flow of the Thames at Isleworth in west London was only a quarter of its summer average, while its source had dried up for 14 km. Dead fish littered the cracked and dried-out lakebeds of Wales. Some trout fisheries lost their entire stock and distressed fish were to be found in canals, lakes and rivers in many parts of the country. Most were suffocating from a lack of oxygen in abnormally high water temperatures; some were unable to survive in increasingly polluted, sluggish water. Angling clubs and water authorities were advising the nation's three million freshwater fishermen to stop. Even cricket might have welcomed a little rain. Bowling averages became higher and higher as wickets got slower and outfields harder. (No one bothered to tell Michael Holding this. In the final Test at The Oval, the West Indies opening bowler achieved figures of 14 for 149, on what the *Daily Telegraph* correspondent described as 'a dead pitch, at the sight of which most fast bowlers would have winced'.) There was said to be a grim race on between the end of the season and the breakdown of pitches. As cattle were dying in the West Country and the inhabitants of south Wales were preparing for 17-hour a day cuts to their water supplies, it was little comfort that sportsmen in the Highlands of Scotland could expect record numbers of grouse during the forthcoming season, after successive mild winters, dry springs and a good summer.

When the emergency legislation came into force on 5 August, three water authorities in Yorkshire, Wales and the South-west immediately sought to cut off water at any time. In Scotland, outwith the provision of the Drought Bill, there were suggestions that fleets of water tankers would soon be heading south and that the Royal Navy might be called upon to deliver water to the stricken South-west.

Seven other authorities introduced a ban on fountains and the watering of parks, gardens, recreation grounds and sports fields. With the imminent start of the rugby and football seasons, the impact of the drought on sport was about to increase drastically. Playing fields and football pitches, together with athletics tracks, tennis courts and municipal bowling greens, all became subject to severe restrictions. In Norwich, council-owned pitches were closed until 1 January. In many London boroughs, water supplies to football pitches had stopped by the end of August. With scorched grass and wide cracks in abundance, there was hardly a playable surface among the 300 pitches run by the Greater London Council, and amateur clubs were left pondering when, or if, their season could start. Envious eyes were cast on Hackney, the proud possessor of an Astroturf pitch.

Sports administrators suddenly woke up to the crisis. The RFU issued a statement warning clubs to consider carefully whether or not to play matches under the current bone-hard conditions. It was probably a bit late to think about this on 24 August – London Scottish had already been training twice a week for four weeks on a grassless, rock-hard pitch at Richmond. No injuries were reported . . . yet. The RFU suggested that games should only be played where there was 'a reasonable covering of grass', but would neither prevent clubs from playing nor alter the start date for the season, 4 September. They did insist, however, that Harlequins find an alternative venue for their matches, as Twickenham was out of bounds until there was rain. The Welsh Rugby Union left a decision to the discretion of the clubs but also advised against play until it rained. The Rugby Football League showed more initiative. Recognising that grounds were approaching a dangerous level of firmness, they advised clubs to negotiate with public health authorities about the possibility of using purified sewage water for pitches.

So began one of the more bizarre solutions to an uncommon British weather problem. New Hunslet became the first club to use impure water on the ground at Elland Road Greyhound Stadium in advance of their Yorkshire Cup tie against Keighley. The water, obtained from a local vehicle body building company, contained oil and other waste materials but tests revealed that it posed no health hazard. Presumably the main concern was for players and groundstaff, but the health of the pitch also had to be considered. Grass can react badly to chemical substances, as the groundsmen at the Hurlingham Cricket Club had discovered. They wondered why the grass on their field died in spite of thousands of litres of water being applied over the summer. The source of supply was the Thames at Fulham but, with a much-reduced downward flow of fresh water, the river had become increasingly salty as the percentage of tidal water increased. Grass does not like salt: the result was lethal.

Football clubs soon took up the idea of liquid refuse. Torquay United went ahead with a scheme to use waste water from a local sewage works, but Exeter City had second thoughts about pouring 46,000 litres of treated sewage effluent onto their pitch until the health risk had been assessed. London sports grounds only received the go-ahead to use effluent on 28 August, the day the drought broke in the capital after five and a half rain-free weeks. Three days earlier, a Minister for Drought had been appointed; it was none other than Dennis Howell, Minister for Sport.

Meanwhile, horse-racing authorities had got to grips with the issue by devising a compensation package for any racecourse forced to transfer a meeting to a different venue because of hard ground. Epsom had taken advantage of the scheme by shifting its August bank holiday meeting to Kempton. The Horserace Betting Levy Board reiterated its policy of maintaining racing to ensure at least two meetings each day and three on Saturday, but announced 'a vigorous cutback' in the number of fixtures. From the end of August to October, at least 19 National Hunt meetings would be abandoned, mostly in the West Country, unless there was a dramatic change in the weather. A delegation from the racing industry, including representatives from the Jockey Club, the Racecourse Association and Newbury, Ascot and Kempton racecourses, approached Thames Water Authority, requesting permission to continue watering two of the three regional courses to enable racing to take place. Cheltenham racecourse upset Severn–Trent Water Authority and members of the public by watering sections of relaid turf without which, according to the course management, the National Hunt Festival in March would be in jeopardy. It nearly was, but for an entirely different reason, as we shall see . . .

Other sports found alternative solutions to the immediate problem. The polo season came to an end with a match at Cowdray Park, Sussex, made possible by the application of local river water. Kelso Rugby Club took the precaution of watering the pitch ahead of its centenary seven-a-side tournament, by pumping water from the River Tweed. Cardiff had sunk boreholes around the Arms Park ground and Harlequins had arranged to play their first four home games at the Stoop Memorial Ground. In the worst hit area, south Wales, the local water authority promised that supplies would be maintained to the St Pierre Golf Club, near Chepstow, ahead of the Dunlop Masters Invitational, in spite of severe local water cuts for domestic consumers. Football writers, noting a string of pre-season injuries amongst First Division players, gloomily predicted a lot more if the drought continued.

But of course it didn't. It began to rain during the August bank-holiday weekend, a piece of bad timing for a workforce that had sweated its way through the long summer months in overheated factories, shops and offices. Eight county cricket matches and the one-day international between England and the West Indies at Lord's were all interrupted by rain. At Folkestone, the pitch was waterlogged and there was no play. At Trent Bridge, there were seven stoppages. Once it had started, it just kept on raining throughout September and much of October. The Kelso Sevens took place in driving wind and heavy showers, and the crowd was one of the lowest ever at the event. There was almost continuous rain at Chepstow for the Dunlop Masters which 'made the mud glisten'. It was the wettest September since 1918 and the total rainfall from mid-September to mid-October was the highest then recorded, at 172 mm. There was more egg on the faces of the Met Office personnel, who had issued a special autumn bulletin giving little hope of sufficient rain and the need for 'two nice, wet winters'. The forecasters, basing their long-range predictions on previous patterns for August and September, had got it spectacularly wrong as the long, hot summer quickly deteriorated into a wet, muddy autumn.

A rapid change took place throughout the country, but it was most noticeable in those areas in which water rationing had been enforced. Conditions for horse-racing in particular altered significantly. National Hunt racing at Devon & Exeter had been cancelled in August as water supplies in the county had been cut off and standpipes erected in mid-September. Two weeks later, on 26 September, the official state of the going at Newton Abbot racecourse, near Torquay, was described as soft. By the following day, it was heavy. At Fontwell Park racecourse near the Sussex coast, another area of water restriction, the ground was so hard in mid-August that two meetings had to be abandoned, and it was still hard on 14 September. Then 25 mm of rain fell in less than an hour and roads in the vicinity were suddenly 300 mm deep in floodwater. A week later, at the race meeting at Lingfield Park, only 64 kilometres away, it rained so heavily that one of Pat Eddery's mounts trailed in last because, according to the jockey, 'he simply could not pull his feet out of the ground.' Newspaper tipsters began to warn punters to be cautious in selecting their bets as the frequency and intensity of downpours caused the ground conditions to change extremely fast. Horses that were proven mudlarks were very much in favour.

The wet autumn had started even earlier in the North. On Friday, 10 September, the going at Sedgefield, near Middlesbrough, was officially described as hard. On Monday 13th, the meeting was abandoned because water was lying on parts of the track after heavy rain and high winds. The local football derby between Middlesbrough and Sunderland, played in torrential rain at the weekend, had seen touch-line markings washed away and groundstaff, muffled in oilskins, attempting to repaint them several times during the match. A public announcement urged a spectator to return immediately to his Teesside home as it was threatened by floodwater. Further north, the Home International golf tournament at the Muirfield course, East Lothian, was disrupted by rain and strong winds; further south, the world championship yachting at Weymouth was halted by a northerly gale, and gusty winds upset the horses competing in the dressage phase of the Burghley Horse Trials.

As a soggy September wore on, mutterings of discontent began to surface in the press and on the streets. Those in areas of water rationing such as Devon, south Wales and West Yorkshire started to question why it was necessary when flooding and heavy rain now seemed to be prevalent. New restrictions were even introduced. In Renfrewshire, hosepipes were banned for the first time; in the Anglia water area, a ban on the watering of all sports facilities came into operation. A verbal tussle took place in the realm of officialdom. Water authorities continued to issue dire warnings, citing cumulative rainfall deficit since May 1975 and low average rainfall

LEFT: Touchline, what touchline? (Sports Turf Research Institute)

BELOW: A bit more than casual water: not the Swilcan Burn but the aftermath of a deluge at St Andrews.
(Cowie Collection, St Andrews University Library)

Who would be a jockey? Snow at Sandown.
(Gerry Cranham)

TOP: No dress code here: the heat is on at Bold Miners' Welfare Bowling Club.
(Ian Beesley)
ABOVE: Who cleans their kit? A less glamorous side to Premiership goalkeeping.
(Ian Beesley)

RIGHT: The weather wins again.

BELOW: Not a day for the fancy
dan.
(Sports Turf Research Institute)

Getting on to the slopes is sometimes harder than getting down them.
(CairnGorm Mountain)

ABOVE: It also happens elsewhere: South Africa v England at Centurion Park, Verwoerdburg, 1995.
(Patrick Eagar Photography)

'Would you rather watch the rain at Wimbledon or the rain at Lords?'

(Telegraph Group Ltd, 1991)

TOP: Snowed off: Saturday afternoon outside a deserted Valley Parade, Bradford.
(Ian Beesley)

ABOVE: Widnes coaches trying to keep dry at Valley Parade.
(Ian Beesley)

TOP: 'High' tech: the funicular ascending Cairngorm.
(CairnGorm Mountain)
ABOVE: A roof with a view: the Millennium Stadium, Cardifff.
(Huw Evans Picture Agency)

figures between June and August of the current year. Meteorologists countered with soil moisture deficit figures, which were said to be a more accurate measurement, and suggested that the situation was not as critical as predicted. The Drought Minister said that some kind of system for moving water from areas of plenty to those of scarcity should be developed, and agreed that the Drought Bill should have been brought in several weeks earlier. The chairman of the National Water Council rejected the concept of a national water grid as 'not feasible'. To cap it all, the Met Office put out its latest long-range forecast in the middle of the month, predicting a good deal of dry, settled weather, broken from time to time by rainy spells.

Presumably no one was looking out of the window! It continued to rain. At the beginning of October, gloomy forecasts about lack of grass were confounded as there was now enough for sheep to graze in some areas. By mid-month, nearly all restrictions had been lifted in the South-west water area, and a report by the National Water Council sounded a relatively optimistic note about future water supply – this from a body whose earlier warnings suggested that it could take two winters and a normal summer to restore reservoirs to proper levels. Only four regions were still operating restrictions at the end of October, by which time Brough Scott, writing in *The Times*, likened the ground at some racecourses to 'a farm gateway in February'. Driving rain and wind continued to, as one correspondent put it, 'reduce the value of the more subtle footballing skills'. By early November, football matches were being played on pitches ranging from 'soft' (Arsenal) through 'yielding to wet' (Ipswich) to 'heavy' (QPR). In December, one of the more imaginative sports writers described West Ham's Upton Park as 'a Christmas pudding of a pitch'. Fanciful perhaps, but we get the general idea – football grounds were back to normal.

And so was the weather – and many of the reservoirs, some of which were full to overflowing by February. In March 1977, the government announced the setting up of a National Water Authority, with wide-ranging powers, to replace the National Water Council but, by then, the summer of 1976 seemed like a distant dream to the majority of Britons. A wet autumn had been followed by a premature winter. The Pools Panel sat on 5 December for the first time in three years, and conditions were so bad at Murrayfield that the underground heating system failed to cope with snow for the first time since 1959. As for the National Hunt Festival at Cheltenham, it took place not, as predicted in August, on rock-hard turf but in a sea of mud, and it was lucky to be held at all. The 1976–77 National Hunt season turned out to be one of the worst on record for abandonments. Meetings were lost initially to hard ground, then to waterlogging, then to frost and snow. The flagship fixture at Cheltenham was viewed by some as 'a shimmering oasis in the race-drained desert of the present season'. On the opening day, the wind howled and rain pelted down – conditions were described as almost intolerable.

Thus ended the great British drought, wrongly predicted as a harbinger of worse to come. In its aftermath, there were massive problems for householders whose property stood on clay soils, now baked dry and liable to subsidence. For sport, there had been some temporary inconvenience, dislocation and financial cost. The chief legacy of the 1976 drought, however, seemed to be that, for a few years at least, Britain developed a different attitude to water, seeing it for the first time in the way other countries do, as a precious commodity.

BEST OF BRITISH

11

A Place in the Sun: Wembley Cup Finals

Planning a wedding? A barbecue? A spring fair? In fact, any event for which sunshine would be welcome and rain might spoil the day? Before you fix the date, find out when the FA Cup final is to be played. That way, you have a 70 per cent chance of sunny periods and an 80 per cent likelihood of dry weather. It is incredible but true that the showpiece of the English football season, normally played in May, seldom takes place in rain, even though the opening of the cricket season is frequently cold and wet. The Wembley pitch may have been heavy, damp or slippery, and the football dull but the sun usually shone on the festivities.

Over the past 50 years, from Matthews to the Millennium, good weather has predominated. The 1950s were particularly blessed. Match reports began to refer to typically blue Cup-final skies, and 'the proverbial Wembley weather', meaning glorious spring afternoons. The turf was always compared to a freshly mown lawn, although in 1954, when West Bromwich Albion beat Preston 3–2, the lack of rain had left it slightly tinged with brown. The problem in the 1950s and early 1960s was not the playing conditions but the injury jinx that settled over the match. Teams were reduced to ten men at least six times in a decade, either through accidents or foul play, at a time when substitutes were not allowed. Nottingham Forest were the only side to win the Cup while playing a man short, against the luckless Luton Town in 1959. (One of the authors allegedly wept buckets on this sad occasion for her local team and threw away her black and white rosette in disgust.)

Weatherwise, the 1960s were less fortunate. Wolves were beaten 3–0 by Blackburn on a warm day in 1960 when the temperature reached 22°C and the crowd, mostly in shirtsleeves, resembled spectators at a midsummer cricket match. Thereafter, the Cup final was played in mixed weather. There were strong winds and showers when Spurs won the first of their consecutive finals in 1961 against Leicester, and drizzle when they beat Burnley the following year. There was heavy rain during the 1965 match – Liverpool triumphed 2–1 over Leeds in extra time – and more blustery showers when Spurs won for the third time in seven years in 1967. The last finals of the decade saw damp, grey days with football to match, according to the soccer scribes. One goal was enough on each occasion to give West Brom and Manchester City wins over Everton and Leicester.

The 1970s started inauspiciously but got better and better. The spring of 1970 was poor with rain, sleet and snow, and the Cup final was played almost a month earlier than usual, on 11 April, because of the forthcoming World Cup in Mexico. The heavily sanded pitch was terrible, the result of recent weather, poor drainage and the aftermath of the Royal International Horse Show, staged at Wembley during the wet summer of 1969. Ground conditions were held responsible for two of the goals in the 2–2 draw between Chelsea and Leeds; Chelsea won the replay at Old Trafford. It was grey and wet in 1973 when Sunderland overcame Leeds by a single goal, causing miniature fountains to appear whenever a player struck the ball hard. The scoreline was the same in 1978 when Ipswich beat Arsenal on a pitch that was none too great after a week of rain. The remainder of the decade, however, saw Cup finals played in dazzling sunshine, in sweltering heat and on fine May days. The football seldom outshone the weather, with Geoffrey Green of *The Times* commenting that nervous paralysis seemed to have taken over from the injury jinx as a wrecker of good matches at Wembley.

The 1980s started with a glorious May afternoon that saw West Ham beat Arsenal 1–0. It was so warm that players had to pace themselves carefully in the 'devitalising sunshine'. When Spurs did the FA Cup double again, with consecutive wins in 1981 and 1982 to match their triumphs of 20 years earlier, the weather was mixed and the surface greasy. In 1983, after a wet spring, the pitch was described as a bog and as heavy as wet sand. Brighton, recently relegated from the First Division, nearly beat Manchester United in extra time, but the strength-sapping conditions contributed to a costly miss in the last moments of the game. United won the replay 4–0. Thereafter, it was back to classic Wembley weather. The sun baked down and football correspondents struggled to find enough suitable adjectives to pepper their reports. Warm, sunny, glorious, gorgeous, oppressive, scorching and un-English were trotted out, the last in 1989 when Liverpool beat their Merseyside rivals, Everton, 3–2. More than 250 spectators received medical attention, mostly for heat exhaustion, in the warm, airless stadium; the temperature on the pitch was said to have reached 32 °C. Players tired visibly in the heat and extra time but Liverpool won through in this year of the Hillsborough disaster.

Continental weather did not continue into the 1990s, a grey decade until the last few years. It was grey and still when Spurs repeated their Cup success in 1991 to mirror 1961 and 1981, winning 2–1 against Nottingham Forest. It was grey and wet in 1992 as Sunderland tried to repeat their 1973 success, achieved in similar conditions. The rain stopped just before kick-off, but this time they lost 2–0 to Liverpool on a day when players slipped and skidded on the wet grass. The year 1994 saw relentless drizzle, and 1995 was grey. Where had all that wonderful Wembley weather gone? The old stadium was certainly dripping with sweat in 1997, but it was more sultry than sunny when Chelsea beat Middlesbrough in sticky, stultifying heat. The century ended, however, in a blaze of sunshine. It was once more possible to write that the weather was kind 'as usual' and the new decade has continued in similar vein. It was a gloriously sunny day when Liverpool beat Arsenal 2–1 in 2001 and Arsenal overcame Chelsea 2–0 in 2002 – at the Millennium Stadium, Cardiff.

12

A Walk on the Wild Side: The Open Championship

Carnoustie golf course on the blustery east coast of Scotland has the reputation of being one of the toughest golf courses in Britain, even when the wind is not blowing. At the Open Championship held there in 1999 it did blow – incessantly for the first two days. The greenkeeper at Carnoustie, John Philp, had the view that it was him, God and nature against golfers and their technology, so he kept the fairways narrow and the rough deep. The wind took charge. Too many players were unable or unwilling to adjust their game to the conditions. The scorers had to send out for more blue figures for their board as they ran out of these coloured indicators of bogey scores. Of the 156 starters none broke par on the first round and only one equalled it: the average score was 78.3, more than six over! Australian Rodney Pampling, alone in making 72 on that first day, enjoyed his 18 holes of fame but on the next day could not cope, took 86 and failed to qualify for the final rounds. Even worse was the performance of rising Spanish star Sergio Garcia, who was 30 over for the first two rounds and finished last of those who missed the cut.

The elements, like the contours of the course, are part of the challenge facing the golfer but, apart from pin placement on the green, these physical aspects are held constant during a competition, unlike the weather. Most major tournaments today occupy four days, so different climatic conditions can be faced within the same event. This is often the case with the Open Championship, which always takes place on seaside links courses susceptible to wind shifts and weather changes. In 1981 at Royal St George's in atrocious conditions on the first day, Jack Nicklaus shot 83, one of the worst rounds of his professional career. When the wind abated on the second day he scored a 66. Weather conditions can also vary from year to year. The 1981 Open saw only one player break par for the four rounds. Four years later on the same course in similar weather no one did it. Yet in 1993, when St George's was bathed in glorious sunshine, Australian Greg Norman set an Open record with an aggregate 13 under. But many other players also broke par to reinforce the point that the conditions, good or bad, are the same for all players.

Almost every Open has a weather story attached. In 1996 the benign warm and dry conditions put Royal Lytham and St Anne's at the mercy of the players, with all three par fives reachable with an iron second shot. In 1986 the wind on the first day at Turnberry blew away leading players and saw rounds of 78 from Masters' champion Jack Nicklaus, US Open champion Raymond Floyd, and Open titleholder Sandy Lyle.

Occasionally the players have had to contend with heat and drought. At Royal Birkdale in 1976 the summer's unrelenting dry spell led to tinder-dry conditions. Barrels of water with stirrup pumps attached were scattered around the course and placed close to all stands, because of the danger of fire from dropped cigarettes or broken glass. As it was, play was held up when a fire broke out in the gorse between the first two holes. Scorching weather had also made the greens patchy and inconsistent with even short putts causing problems. Yet, as often happens at the Open, the weather changed. On the third day the wind picked up and the rain joined in so that there was only one sub-70 round. None of this stopped Johnny Miller becoming the 13th golfer to win Opens on both sides of the Atlantic, but defending champion Tom Watson took an 80 and failed to qualify for the final day. In the heat wave at Royal Troon in 1989 players found the ball travelling much further than expected because of bounce on the baked fairways. This proved disastrous for Australian Greg Norman who, on the fourth and final hole of the play-off, found his booming drive ricocheting into a fairway bunker, previously thought out of reach of any tee shot. Norman's failure to reach the green with his recovery shot – he actually went straight into another bunker – effectively presented the title to American Mark Calcavecchia. Three years later at Muirfield the players appreciated a course that had scarcely seen a drop of rain for two months and there was a glut of scores in the sixties.

More often the weather has been wet. Henry Cotton scored an acclaimed 71 in the final round of the 1937 Open Championship at Carnoustie to become the first British player to win an Open when the field included the whole American Ryder Cup team. At one under par the score might not seem astounding, but it was achieved in blinding rain that began early in the morning and never stopped. Local shops made a killing that day selling waterproofs for spectators. The start was delayed for half an hour while groundsmen swept water from the flooded first green and members of the championship committee were despatched to distant parts of the course to assess whether play was possible. Players had to move their ball for virtually every putt so as to gain a clear water-free line to the hole.

In addition, there has almost always been the wind: crosswinds, headwinds and downwinds, all contributing to the uniqueness of Open conditions. Take St Andrews as an example. In 1900 a strong easterly breeze made for testing conditions. Three years later a cold strong wind prevented any competitor breaking 80 on either of the first two rounds though, of course, these were the days before titanium, Big Berthas and hollow-backed irons. Thirty years later *The Guardian* reported that 'a fierce wind' sprang up from the sea and made the inward nine 'exceptionally difficult'. It was still a 'troublesome wind' next time the championship came to the Old Course in 1939. By 1970 it was 'violent to gale force in gusts'. In between it had 'not made things easy' in 1964. The last time St Andrews hosted the Open, in 2000, Jack Nicklaus was heard to comment that to win there a golfer 'must learn what those ever-changing winds will do to his ball, relative to a seemingly infinite number of invisible or partially hidden hazards on each and every shot he might be called on to play – including sometimes putts'. And Jack loves the course! But it is not just at St Andrews that the wind can be difficult. Harry Vardon reckoned that the 1907 Open at Hoylake, won by Frenchman Arnaud Massy, had the worst wind that he ever experienced. The short fourth hole required a full driver shot and even that had to be kept low to escape the full force of the gale raging overhead.

Vardon died in 1937; otherwise he might have revised his opinion. The 1938 Open Championship at Sandwich was fated. It should have been held at nearby Deal, but had been transferred when storm damage and flooding hit the original venue. On the eve of the final day a hospitality marquee was ripped to shreds in high winds. It was a harbinger of what was to come. On the day itself, a gale swept across the course, rendering distance calculations superfluous and making it difficult for players to keep their balance, especially on their downswing. With the wind, the 11th hole of nearly 410 yards could be reached in one shot, though with little control over where the ball might finish. Against the wind, one competitor took 14 strokes at the 12th, including 3 out of bounds, and the 520-yard 14th was almost unreachable even with three full woods. On the greens the balls quivered on the putting surface, aggravating the difficulties of players already struggling to balance themselves in the ferocious gusts. Of the 37 qualifiers for the final rounds, only 9 broke 80 in the morning and a mere seven in the afternoon. The winner, Reg Whitcombe, the youngest of three brothers, all of them Ryder Cup players, did it both times. In more recent years the wind at Royal Troon in 1997 meant that players had to abandon their yardage charts and instead imagine their shots, playing more by feel than by automatic pilot. The following year at Royal Birkdale, Sir Michael Bonallack, head of the championship committee, acknowledged that at times 'the wind was just too strong for it to be a fair test', with some drives not even reaching the fairway.

Sometimes the climate has thrown almost everything at the players. Ben Hogan's win at Carnoustie in 1953 was accompanied at various stages by rain – slight, heavy and torrential – grey skies, hail, mist, a fierce wind and, to complete the meteorological set, occasional sunshine. Despite this, Hogan slashed eight shots from the record Open aggregate at the Angus course to become the first player to win the championship at his first attempt since Willie Park at the inaugural Open in 1860! Carnoustie's climatic conundrums were solved easily by a man who only four years before had lain in intensive care following a horrific car smash that fractured his pelvis and mangled his left leg.

Play has been delayed, suspended or even abandoned on several occasions because of the weather conditions. In 1910, when James Braid won his fifth Open (at St Andrews) the first round was abandoned at 1.30 p.m. because of a thunderstorm, a day added to the schedule and, a novelty in those pre-cut days, only the 60 leading players were allowed to play the final two rounds. In 1960, again at St Andrews, the heavens opened early in the final round. The well-travelled Arnold Palmer exclaimed that, 'It rained like I've never seen it rain before, coming in wild gusts and torrents.' Water poured down the steps of the Royal and Ancient clubhouse; the Valley of Sin, a landmark depression approaching the final green, became a miniature lake; and the Old Course became unplayable. The round was called off and Ken Nagle had to wait another day to secure the championship in its centenary year.

In 1961 at Royal Birkdale a monsoon – or so it seemed – washed away the tented village and forced play to be abandoned for most of the third day. The wild weather had already caused substantial damage to the marquees and temporary buildings on the opening day. Arnold Palmer, the eventual winner, donned several sweaters and a cap to face the driving rain and howling wind. The wind was so strong that he needed a 1-iron to reach the 212-yard fourth and a driver and a 6-iron at the 315-yard fifth. He carded a 73 in what he rated one of his finest rounds. At one stage the championship committee threatened to call the event off if play was delayed too long. The rain relented: a miracle, or just typical British weather?

After the start had been delayed by mist at St Andrews in 1970, defending champion Tony Jacklin shot seven under par for the outward half of his first round, but a cloudburst forced play to be abandoned for the day as he reached the 14th. The spell was broken and on resumption the next morning he dropped shots at two of the final holes. Eventually he finished fifth, three shots behind Jack Nicklaus and Doug Sanders, but who knows what he might have achieved had weather not interrupted his hot streak.

The Open at Royal Lytham in July 1988 endured 22 hours of unremitting rain that drenched over 35,000 spectators (a misnomer, as there was nothing to see), washed out the Saturday and led to the first Monday finish and a third championship for Seve Ballesteros. No equipment could have coped with the deluge, but there was something quintessentially British about the greenkeepers ridding the greens of excess water by means of inverted rakes. A decade later at Royal Birkdale the second day's play was suspended in the late afternoon after lightning was discerned over the Irish Sea. When play was resumed, an 80 kph wind swept in to buffet the players, bend the flagsticks and move balls supposedly at rest. For once, more lightning – for which play is always called off – might have been appreciated on that brutal evening.

Links courses need foul weather to protect them from the modern professional golfer and his technological armoury. When there is wind and rain the links test the players as much as their play. The weather can force them to manufacture shots that they would never practise and in tough conditions it is the mentally strong who will come through. Those who adjust their tactics and make the right decisions will do better than those who merely complain and do not change their game. On the day, it is the competitor who copes best mentally and physically with the climatic circumstances who will triumph. Just think of Tom Watson, who beat the final day's strong wind, dripping rain, severe pin placements and Jack Nicklaus in a play-off on his first attempt at Carnoustie in 1975, or Nick Faldo, who brightened gloomy skies for home supporters when his new swing overcame a sodden course and a gale-force north-easterly wind to produce victory at Muirfield in 1987. Simply put, it is up to the golfers themselves to deal with what nature throws at them.

13

Service Interrupted: Wimbledon

'Ladies and Gentlemen, Play is Suspended'

It's early evening on Friday, 6 July 2001 in SW19. British radio and television commentators can scarcely conceal their excitement. Newspaper editors are holding the front page and preparing to issue special weekend supplements, waiting, waiting . . . for, within the next half-hour, history might be made. The fervent Centre Court crowd makes no attempt to contain its feelings, cheering and shouting, flag waving and holding its breath. And then, at 6.18 p.m., it begins to rain. The sense of dismay, frustration and anti-climax as Tim Henman and Goran Ivanisevic retreat to the locker room is almost overwhelming. Why, oh why, have the rain gods seen fit once again to cheat the Wimbledon tennis crowds of their long cherished dream – to see a British men's singles finalist? They wait patiently, but in vain. The presses roll, without those longed for banner headlines – 'Tim Reaches Final', ' He's Done It', 'Hero Henman – First British Men's Finalist since 1938'.

Tim Henman, having played possibly the best set of his career, was leading 5–7 7–6 6–0 against three-time losing finalist Goran Ivanisevic in the second semi-final of the day. At 2–1 in the fourth set, most observers reckoned that he was on his way to glory. But the weather can be cruel. It was hard enough to abandon the match on Friday evening and return the next day; harder still to sit around all Saturday as a light but steady drizzle continued to fall. The players emerged just after 5.30 p.m. when it finally stopped. An hour later they were back indoors with Henman having lost the fourth set, despite leading 3–1 in the tie-break, and now ominously 3–2 down in the fifth. Another day of disappointment and frustration, another night of anxiety for players, press and public.

Finally, everyone was put out of their misery. Forty-five hours and nine minutes of rain and intermittent tennis eventually produced a result, but not the one the Centre Court had hoped for. The editors put away their headlines for another year, the Union flags returned to the closets and Tim Henman bowed out of

Wimbledon, beaten by nerves, rain and his Croatian opponent, who wrapped up the final set in less than 30 minutes on Sunday afternoon. There was a real sense of dismay that the fickle British weather had contributed so greatly to the downfall of the nation's number one tennis player.

There were recriminations. Why were Andre Agassi and Pat Rafter allowed to play their semi-final first? Why did play start at 1 p.m. when it was known that rain was likely to intervene before both matches were completed? And, for the umpteenth time, why doesn't Wimbledon have a retractable roof which would have prevented the whole fiasco? Why indeed.

The Retractable Roof Saga

When the rain falls pitilessly on Wimbledon, as it does most years, the press and public demand answers to an age-old question: why isn't there a movable roof over the Centre Court? It is unlikely, after all, that an average summer in London will be completely rain-free for two weeks. The same question has been asked repeatedly for over 30 years, but the answers have varied. At first, it was considered to be an architectural and engineering impossibility. The existing partial roof was not strong enough to support a rigid sliding structure and all the roof column supports would have had to be rebuilt over a period of two years at a cost of 'tens of millions'. In 1988 an alternative lighter covering was said to be impractical because it could be damaged by heavy rain and high winds, and was as dangerous as 'a runaway umbrella', according to the structural engineer on the Wimbledon management committee. However, one firm of American architects and engineers predicted minimal disruption, a time scale of less than 12 months to completion and an estimated cost of around £3 million. (£4.8 million today or, at that time, one and a half Paul Gascoignes!) They suggested a concertina structure of PVC-coated polyester, which could be folded back when not required, and which, if properly constructed, was perfectly safe. Nothing was done.

In 1991 an alternative suggestion costing around £10 million was put forward. It would have involved suspending a retractable roof from masts, allowing not only the removal of the existing framework and pillars but the circulation of air around the stadium. This time a spokesperson for the All England Club made the excuse that noise would be a problem as it rebounded from the ceiling, and that any machinery which gave off gases, such as generators, could not be used when it was closed. In fact, the most telling part of this statement concerned the main objection: that a roof would 'completely and utterly change the nature of the event'. Setting aside the issue of maintaining the hallowed grass of the Centre Court, what this really meant was that 'tennis in an English garden', or, as one correspondent put it in 1975, 'that engaging annual garden party of the tennis world' would not be sacrificed in the name of expediency. After more than 20 years of open, professional tennis, players and public would still be required to suffer the inconvenience of regular rain breaks and soakings in order to maintain an atmosphere redolent of an Edwardian summer afternoon.

This attitude was confirmed in 1993 when the blueprint for the redevelopment of No. 1 Court, due for completion in 1997, was unveiled. The brand new all-seater court, with nearly double the previous spectator capacity, was designed without a retractable roof, although the structure has the strength to support one if required. A major argument against a covered court had always been the difficulty of attaching a sliding contraption to an existing frame. Now, when the opportunity of building from scratch was available, the option of weatherproofing at least one Wimbledon show court was rejected.

By this time, there were precedents. The sliding roof at the Flinders Park centre court in Melbourne had been operational for several years, albeit over a hard court surface. Closer to home, and more comparable, was the new Perspex roof at the Halle stadium in Germany, constructed in 1993–94 and unveiled in time for the 1994 men's grass court tournament in June. On a damp and drizzly day, Jim Courier and Bernd Karbacher became the first players to perform on covered grass. The Halle event, known as the Gerry Weber Open after its founder, is now used as grass court preparation for Wimbledon, although the German tennis complex originally had clay and concrete surfaces. When Weber decided to apply for a mainstream ATP event, he was offered a slot two weeks before the British Grand Slam tournament. Under these circumstances, he brought in a former All England Club groundsman, converted the courts to grass and built a 12,000-seat centre court with a sliding roof, which closes in 88 seconds. A delegation from Wimbledon, including the present head groundsman, is said to have visited Germany to examine the facilities at Halle. The benefits should have been obvious in June 2001; when play was washed out at the Stella Artois tournament in London, the final of the Gerry Weber Open was completed in spite of torrential rain.

The grass itself has now become the main reason why Wimbledon apparently can't have a roof. In 2002, old arguments about the impossibility of construction have been dropped, with officials conceding that it is perfectly feasible. 'Microclimate' has become the buzzword instead. With the pitch problems experienced at Cardiff's Millennium Stadium, which possesses a retractable roof, and at football stadiums such as Manchester United's Old Trafford, which does not, the issue of natural surfaces in the modern sports venue has never been more topical. Wimbledon officials now accept that experiments taking place in America should help to provide answers to several important questions: will a grass court sweat under a roof, posing a threat to player safety; how will grass growth be affected by roof closure; what will be the impact on playing conditions and spectator comfort? Until these difficulties, together with scepticism amongst some players, can be overcome, there will be no alternative to the well-drilled rain-covers teams at courtside.

To be fair to Wimbledon, other avenues have been explored in recent years and different solutions to the weather problem advanced. Play has been brought forward from 2 p.m., with the exception of finals weekend, to improve light and reduce the necessity for twilight matches on dew-damp courts. (They could, of course, start at 11 a.m. like other major tournaments!) Pinpoint accuracy of short-term weather forecasts supplied by the All England Club's radar systems, and the use of palletised turf technology, due for trial in 2002, should provide additional benefits in the constant battle to keep the championships ahead of the elements. Unfortunately, the days of sodden spectators are not over.

GREEN, GREEN GRASS

Rain breaks can turn a match upside down; swirling winds can adversely affect the big servers in particular; too much heat can cause dehydration. In addition to weather problems, Wimbledon contestants have to deal with grass, a very different environment from clay and the man-made hard courts. It is the fastest and most unpredictable surface and, because of the weather, the least consistent over the period of a championship. If there is insufficient sunny weather before the tournament, the courts will not harden up enough to provide the high bounces favoured by baseliners. If it starts to deteriorate during a sun-baked fortnight, little can be done to slow the process, although Wimbledon courts are sown with a high percentage of hard-wearing rye grass. If the grass is too green the courts will be slow. If it is subject to frequent showers the surface will become damp, greasy and slippery, no matter how fast the covers are brought on. As if the weather is not enemy enough, the players themselves cause unnecessary damage by constant racket abuse, thumping the grass in frustration. Perhaps they are taking revenge on a surface which many find alien and difficult to play on. There is certainly little opportunity to practise traditional lawn tennis at the highest level in the twenty-first century, with only a handful of ATP tournaments sticking to grass, and some players would like to see it abandoned altogether.

Is a Roof Really Necessary?

If the recent example of the Henman semi-final in 2001 is considered together with the men's singles final of 2000 between Pete Sampras and Pat Rafter, completed in almost pitch darkness at nearly 9 p.m., after lengthy rain delays, then the answer is surely yes. But the weather has been disrupting finals for many years. As far back as 1902 rain ruined the women's singles final. Muriel Robb and Chattie Sterry (née Cooper, the Olympic champion) were one set all when rain intervened and they had to replay from the start the following day: Robb won. In 1963, after a fortnight of persistent rain, the decider between Margaret Court and Billy Jean Moffitt (later King) was postponed till Monday because of the weather. In 1988, the final between Boris Becker and Stefan Edberg was interrupted by heavy rain after only 22 minutes had been played, and became the first men's title in 66 years to be decided on a Monday. The match between Richard Krajicek and Mal Washington in 1996 was also disrupted three times by rain.

The Wimbledon committee has argued that the championships are not about a single court. With 650 matches to be decided, the early rounds can be thrown into chaos by wet weather, a situation that cannot be overcome by one retractable roof. In 1922, the tournament stretched into a third week. Wimbledon 1982 was one of the wettest on record, with over 100 games lost to the elements over the first four days. It rained every day for six days in 1985. By the second week, play was starting at noon instead of 2 p.m. in an effort to make up the backlog of matches. Rain wrecked the first half of Wimbledon fortnight in 1987, wiping out the first day completely, rendering practice courts unusable, cutting attendance figures and turning car parks into quagmires. Wimbledon 1991 was so wet that the first round was not completed until 7 p.m. on the middle Saturday, and the committee reluctantly allowed play on a Sunday for the first time in the history of the championships. This break with tradition had to be repeated in 1997 when two full days were abandoned in the first week, the only time consecutive days had been lost since 1909.

The consequences of rain breaks and delays can seem unfair, although the pundits tend to observe that you're not a worthy Wimbledon champion unless you've dealt with the vagaries of the weather. In 1989, Ivan Lendl would almost certainly have beaten Boris Becker in the semi-finals, according to the tennis correspondent of *The Guardian*, but for a rain break that disturbed his concentration. In the women's singles final of 1992, Monica Seles, a first-time Wimbledon finalist, was undoubtedly upset by a succession of stoppages for rain which contributed to her defeat by old hand Steffi Graf. And nearly 30 years before Tim Henman's traumatic semi-final exit, Roger Taylor, the British number one male tennis player of his era, was left to rue a break for bad weather. In his third and final appearance in the last four in 1973, against Jan Kodes, rain forced the players off court when Taylor was leading 5–4 in the fifth set. When they returned 40 minutes later, Taylor lost both his momentum and the last games, bowing out 9–8 7–9 7–5 4–6 7–5.

Weather breaks balance out over a number of matches according to Alan Mills, the Wimbledon tournament referee. He reminded Henman fans that bad light actually kept their hero in with a chance during his fourth-round tie against Todd Martin in 2001 when he was losing two sets to one. Henman returned the following day to win in five sets on his way to that ill-fated semi-final. The bottom line is that top tennis players, like championship golfers, have got to learn how to deal with the daily problems of life on the tour – the hostile, partisan, inconsiderate or just over-exuberant crowds, the condition of the surface, the hanging around and the disruption and difficulties of playing a game outdoors in all kinds of weather.

Everything but the Kitchen Sink
Although rain is a standing joke at Wimbledon (during the woeful fortnight of 1997 when middle Sunday had to be used again, some wit allegedly cried, 'Let's

ship this tournament to Florida!') less than 30 full days of tennis have been lost to rain since 1877. It isn't only the complete wash-outs which are so frustrating, but also the endless loss of hours here and there, the up and down of umbrellas, the on and off the courts, the reorganising, the placating, the staccato nature of a bad day in SW19. But, as with weather everywhere, it can change rapidly. A day for tumble dryers can be followed by a day for sun block . . . or even 13 days! Amazingly, it can happen. It was hot and rain-free throughout the tournament in both 1976 and 1993. There was virtually no rain in 1977 either, although it was much cooler, and it was so chilly in 1980 that play was interrupted by hail. In 1985, American Anne White appeared on court clothed from neck to ankle in a white ski-style body stocking, claiming that it was too cold for regular tennis clothes. Although she defended the outfit as 'functional' and perhaps the tennis attire of the future, she was asked not to wear it again – at least, not at the All England Club. The same year saw thunder and lightning, which knocked out the electronic scoreboards on the show courts, and torrential downpours, which caused flooding. In 1990, the absence of rain during the first week paradoxically led to the tournament running ahead of schedule, with no play on several outside courts. Wimbledon has it all!

Sometimes it even basks in a heat wave. In stifling conditions in 1952, over 200 spectators were given first aid treatment. Doris Hart and Pat Todd played to the point of collapse over nearly two hours in a women's singles match during the second week. They both ended up on the massage tables recovering from exhaustion, Todd having finally won 6–8 7–5 6–4. She reached the semi-finals, while Hart went on to win both doubles championships with fellow American Shirley Fry and Australian Frank Sedgman. The heat wave of 1989 saw temperatures approaching 38°C on Centre Court, although admittedly the thermometer was out in the sun rather than in the enclosed box of an official weather station. The record air temperature in Britain, taken from such a legitimate source, is still reckoned to be 37.1°C at Cheltenham in August 1990. However, such niceties mean little to players and spectators subject to the heat of an enclosed and sometimes windless arena. The men's final in 1993 between Pete Sampras and Jim Courier was said to have taken place in a temperature of 41°C, the culmination of a fortnight of good weather. On a single day in June 1994, 200 were treated for sunstroke and dehydration . . . and 2,000 bottles of champagne were sold by 4 p.m. (are the two connected?). The following day, with the temperature allegedly over 38°C, Kenneth Carlsen had to withdraw halfway through his third round match suffering from heat exhaustion: he had already been weakened by a stomach upset. Dozens of spectators collapsed, some had to be removed from court on stretchers and the St John Ambulance Brigade warned against too much exposure to the sun.

Heat can upset not just players and spectators but also equipment. Racket strings can be affected by temperature. Balls have to be stored in a fridge during hot weather, otherwise the rubber becomes softer and the balls more bouncy. Instead of protecting the courts from rain, a heat wave can result in groundsmen applying sprinklers each night to keep the grass from wilting and wearing away.

The impact of the changeable British climate on a lengthy sports event such as Wimbledon is enormous. The rain-free tournaments are exceptional, and although bad weather can affect other Grand Slam competitions in Australia, France and the United States, it is Wimbledon that bears the brunt of the jokes. When one of us spent a day at the Australian Open in January 2002, it was hardly necessary to check the weather forecast, even though the climate in Melbourne is probably the least reliable in Australia. Those who book tickets for Wimbledon have to be glued to every available weather report. To sit through five uninterrupted matches on a single day in Melbourne, basking in warm sunshine, is a joy seldom afforded the British tennis fan. Instead, we have to suffer the attentions of Cliff Richard or HM Forces as they lead the inevitable community singsongs and congas during the downpours and drying out periods. Oh, to be in England, now that summer's there!

14

Rain Stopped Play: Test Matches at Old Trafford

In over a century of Test cricket, only five matches worldwide have been abandoned because of adverse weather without a ball being bowled. Two of these were in England, both at Old Trafford, the Manchester home of Lancashire County Cricket Club. Although Manchester itself has an undeserved reputation for wet weather, the same cannot be said for Old Trafford. As the English Test cricket ground with the highest average annual rainfall – at 875 mm, some 275 mm more than Surrey's Oval, which has the driest record – it has an unenviable standing in cricket folklore for weather-affected matches.

This began with the first Test ever scheduled at the Manchester ground in 1884. England were due to play Australia in a three-day Test beginning on 10 July, but the whole of the first day's play was washed out and the game, reduced to only two days, ended in a draw. Two years later play was possible every day and England beat the Australians. In 1888 the rain actually benefited England as it played a major role in their victory over Australia by an innings and 21 runs. Downpours before the match produced a soft, wet pitch that a hot, drying sun made very difficult for those batting on the second day; as the Australians found out. In reply to England's opening 172, Australia resumed their first innings on 32 for 2 and lost 18 wickets before lunch. Their first innings totalled 81, which, under the existing rules, meant that they had to follow on. Their second knock brought them only 70 runs and lasted a mere 69 minutes, their briefest full innings in any Test. The match itself was the shortest Test ever completed. Two years later the third Test, once more against Australia, was totally washed out. So, of the first 12 days of Test cricket scheduled for Old Trafford, four were abandoned without the players taking the field, and a further one was rendered unnecessary because of a rain-assisted victory early on the second day.

Between 1893 and 1909 came six matches against Australia, all unaffected by the weather. Not till 1907, when South Africa toured, did England play a home Test against anyone but Australia. Old Trafford was not awarded a match against the South Africans and Australia continued to be the only foreign visitors to the Mancunian ground until 27 May 1912, when South Africa played in the first Test of a triangular tournament sponsored by Sir Abe Bailey, the South African diamond magnate. Their opponents on this occasion, however, were not England, but once again the Australians. The South Africans proved no match for the Antipodeans in a game uninterrupted by the weather. The Manchester ground had now gone seven matches and nineteen years without rain stopping play. It could not last. When England played Australia in the sixth game of the tournament on 29–31 July, heavy clouds once more rolled in. Rain restricted play to three and three-quarter hours on the first day and one and a half on the second before completely washing out the third.

In the fourth match of the first post-war domestic Test series, played at Old Trafford in July 1921, there was scant regard paid to one law of cricket and a dispute over the strict application of another. When the English captain, the Hon. Lionel Tennyson, attempted to close his team's innings at 5.50 p.m. on the second day, Warwick Armstrong, the combative Australian skipper, all 140 kg of him, objected forcefully. He pointed out that as the first day had been lost to weather the Test was now a two-day affair. Under the rules for such games no declaration was possible unless there was at least 100 minutes batting time remaining on the day. It took almost half an hour to settle the argument and, when the England innings eventually resumed, Armstrong himself bowled the first over, conveniently forgetting that he had also bowled the last one before the break. The match petered out to a draw.

Rain also interfered in the next Manchester Test, versus South Africa in July 1924. A torrential downpour ended the match at 4 p.m. on the first day after only two and three-quarter hours of play. That was the sum total of the Test career of John MacBryan, the Somerset batsman, as he was never again selected. He remains the only Test cricketer never to have batted, bowled or dismissed anyone in the field. The tale of woeful weather continued into 1926 when only ten balls could be sent down on the first afternoon of the fourth Test against Australia. Then came a rain-free match against the West Indies in 1928 and another versus South Africa the following year. Perhaps both opponents wished the heavens had opened, as they lost by an innings and 30 runs and an innings and 32 runs respectively. Normal service – for Old Trafford – was resumed in July 1930 when, in a five-day drawn match against Australia, only 45 minutes play was possible on the third day and none at all on the fourth. And in August 1931, when New Zealand made their Old Trafford debut, a start was not possible until 3.15 p.m. on the third afternoon.

Rain affected none of the next five Tests, yet surprisingly all ended as draws. Even more surprising for Manchester was that, according to *Wisden*, the authoritative voice of cricket, the 1934 match against the Australians saw sunshine 'from first to last . . . the heat being at times almost unbearable'. As if to compensate for this run of uninterrupted games, the third Test against the Australians in 1938 was abandoned without a ball being bowled. So bad were both the wet conditions and the weather forecast that the captains did not bother to toss. There was an improvement for the visit of the West Indies the following year, but only in relative terms. A mere 35 minutes was played on the first day and the start of the second day was delayed. Overall, during the inter-war years Old Trafford lost nine full days of Test cricket, almost twice that of the period before the First World War.

Many things changed in the aftermath of the Second World War. A Labour government was elected; the welfare state began; but at Old Trafford it was the same old story. The opening day of the Test against India in 1946 was delayed until after lunch by rain. In effect, the weather saved India from defeat. Their last two batsmen hung on for the final 13 minutes on the third day, though the English fielders dropped them twice during that tenth-wicket stand.

South Africa were defeated by seven wickets in 1947 on a rain-affected pitch that made later batting difficult. The fourth day of a drawn Test against Australia in 1948 saw no play at all. The next two Tests were untouched by the weather, but in 1951, when South Africa were beaten by nine wickets, there was no play on the second day. After an easy victory against India in 1952, the Coronation year Test against Australia was restricted to only 13 hours and 50 minutes. This was 3 hours and 20 minutes more than against Pakistan the following year, a game in which M.E.Z. Ghazila achieved the quickest pair in Test history, being run out for a duck twice within the space of two hours.

Between 1912 and 1953 Old Trafford had hosted eight Test matches featuring England against Australia. All of them had ended in draws and seven had been affected by rain, two of them so severely that they had been abandoned. One wonders why the Lancastrian spectators kept coming back. Then, in 1956, their perseverance was rewarded by what is conventionally referred to as 'Laker's Test', a match involving one of the most remarkable feats of bowling in international cricket. On a rain-affected pitch that began to spin as it dried, the Surrey off-spin bowler took all ten Australian wickets in the second innings to add to the nine he had captured in the first.

LAKER'S TEST: OLD TRAFFORD 26–31 JULY 1956

The weather was fine for two days: the only storm that broke was over the dustbowl of a pitch that began to crumble early on. Nevertheless England, having won the toss and batted first, managed 459. In reply Australia reached 62–2 at tea on the second day before Jim Laker, the 34-year-old Surrey off-spinner, turning the ball sharply on the dry turf, took 7 wickets for 8 runs in only 22 balls as the visitors collapsed to 84 all out. His overall first innings figures were 16.4–4–37–9. It then looked like rain would come to the aid of the tourists. Only 45 minutes play were possible on the third day and just an hour on the fourth. Although it rained heavily overnight, start of play on the final day was hardly delayed at all, but the pitch was too saturated to give the bowlers any assistance. What England needed was some sunshine to dry the wicket, turn it 'sticky' and let the ball misbehave. At lunch Australia were

112–2, but then blue sky appeared and the warm sun began to work on a surface which had been softened by the previous day's rain. The visitors simply could not cope with the demons of the pitch and Laker's brilliance. After 259 minutes of resistance, Ian Craig played back and was out lbw. Laker now had three wickets. The next three batsmen failed to trouble the scorers. Ken Mackay, known as 'Slasher' because of his ultra-defensive qualities, pushed forward and edged to second slip; all-rounder Keith Miller had his off stump knocked out of the ground; and Ron Archer was caught by the middle of three short legs stationed to take advantage of the vicious spin that Laker was obtaining. He now had six wickets. An early victory and retention of the Ashes looked likely, but opener Ian McDonald and Richie Benaud stood firm, the latter irritating the Manchester crowd with his time-wasting tactics. Every over he asked for a fresh guard and after every ball he expressed a botanical interest in the pitch. The pair survived the 80 minutes till tea. Laker's second ball after the break ended McDonald's marathon 337-minute effort. He pushed forward and Alan Oakman took the catch at backward short leg. Seven down. Benaud, bowled by a generously flighted off-break, was number eight. Ray Lindwall, snapped up in the leg trap, made nine: all of them to Laker. He had already gathered a full complement of Australian wickets in their match against Surrey earlier that summer, when he took all 10 for 88 runs. Could he do it again and become the first man to take all 10 wickets in a Test? He could. At 5.30 p.m. Len Maddocks padded up and was given out lbw. Australia were all out for 205 and lost by an innings and 170 runs. Laker had become the first bowler worldwide to take 19 wickets in any first-class game, let alone a Test match. His second innings figures of 51.2–23–53–10 and his match aggregate of 19 wickets for 90 runs remain the best in Test cricket.

Old Trafford was not allocated a Test in 1957 but in the next four years three of the four Mancunian Tests were affected by rain. New Zealand were beaten by an innings and 13 runs in spite of over eight hours play being lost; the abandonment of the first two days produced a draw against South Africa; and Australia gained their revenge for 1956 by defeating the home side even though half of the first day was lost. This last abandonment of play cracked the century for hours lost in post-war Tests at Manchester.

Of the next five played before 1971, none were affected by the weather. Then in August 1971 there was no play on the fifth day, a matter that probably saved India from defeat. However, the draw was sufficient for England to claim a record of playing 26 consecutive Tests without a loss. Three years later, rain took 5 hours and 38 minutes from the game, but this time it did not save India, who went down by 113 runs. This was not just due to English superiority. New regulations were in place that allowed for an hour's extension on any of the first four days on which at least 60 minutes of play had been lost. In 1980 spectators at a drawn match against the West Indies had nothing to see for 11 hours and 10 minutes, including the whole third day. This was the 25th full day of Test cricket abandoned at Old Trafford, more than twice the total at any other English Test ground. Two years later against India came day 26, giving Old Trafford an average of a full day lost virtually every second Test match: this without counting any stoppages of less than a day! At the time the loss of 26 days in 53 Test matches compared very unfavourably with The Oval (13 in 67), Headingley (12 in 47), Lord's (12 in 77), Trent Bridge (9 in 34) and Edgbaston (2 in 22). Yet attendances continued to justify Tests being awarded to Old Trafford. The Yorkshire half of this writing partnership refrains from comment on the mentality of the Lancashire crowd.

Sometimes the law of averages should not be disparaged. Two Tests against the West Indies (1984) and Australia (1985) were unaffected by the Mancunian climate, so almost inevitably the next, against Pakistan in 1987, was. Just under 14 hours of play was possible and none at all after lunch on the fourth day. The next year England lost the third Test to the West Indies. Had the sun shone they would have been beaten in three and a half days. As it was, they lingered on until lunchtime on the last day, hoping for the intermittent rain to come again and save them. Three minutes after their last wicket fell, so did the rain.

Since 1989, when Australia thrashed England by nine wickets, Old Trafford has hosted nine Tests. Five of these have been affected by the weather, beginning with a drawn game against India in 1990 that was constantly interrupted by showers. So confused did the umpires become with the to-ing and fro-ing from the pavilion that, reminiscent of the Warwick Armstrong incident 70 years earlier, they allowed Anil Kumble to bowl four balls before a break and four balls after it to complete his over. In 1992 the second day of a drawn game, this one against Pakistan, was abandoned at 4 p.m. after persistent rain, the 28th full day's Test cricket to be lost at Old Trafford. However, it was a further five years before wet weather again disrupted a Test at Manchester. Despite three major breaks for rain, England lost to Australia by 268 runs. In 1999 a Manchester downpour prevented possible humiliation at the hands of New Zealand, though such a fate was delayed only until the next Test at The Oval.

All these weather-interrupted games had fallen victim to rain. The other one was surprisingly different. On 28 July 1995 during the Old Trafford Test against the West Indies 'sun stopped play'! Sunlight glinting off the roof of a glasshouse at a B & Q Super Centre at the back of the Stretford End made it impossible for the batsmen to see the ball. At 3.20 p.m., under a cloudless blue sky and to the boos of a capacity crowd, the players were taken off for an early tea. In the meantime, the store manager responded to a call from Old Trafford and allowed the groundstaff to drape the glass with the black cover usually put on the sightscreens when a white ball is being used in one-day games. The interruption did little for the West Indies who were beaten towards the end of the fourth day. The other loser was Nynex, the cable operators, whose name had been emblazoned on the pitch covers in the expectation of television coverage during the rain breaks. Good or bad, the weather in Manchester, as anywhere else in Britain, cannot be guaranteed.

GROUNDS FOR DIVORCE?

15

Roots of the Game

The Aintree Grand National represents either the harbinger of spring or the death throes of winter, depending on how the weather is behaving at the beginning of April. It signals the winding down of top-class steeplechasing in Britain, and the annual stampede to the first fence follows a similar charge through the doors of the betting offices as punters part with over £100 million in bets.

The going is of vital importance for Aintree, as for any turf racecourse. Phil Bull, professional punter and inventor of the Timeform system of rating racehorses, once said that the state of the ground is the most critical element in assessing a race, a variable that greatly affects performance. That variable depends to a large extent on the weather. A wet winter and significant rainfall in the weeks leading up to the big race may cause problems for trainers in preparing horses, but Aintree is a fast-draining course because of its sandy soils; it is a dry spring that brings additional difficulties for groundstaff. Watering a course of two and a quarter miles is a time-consuming job!

Aintree is fortunate in having few race days each year, giving plenty of time for preparation of the track and its 16 fences, which have to be dressed with spruce on the big day. After a field of 40 runners has battled twice round the circuit in weather ranging from warm sunshine to torrential downpour, the damage, particularly to the take-off sides of fences, can be extensive. According to the head groundsman, 'When we come in on Monday morning after the National, it is a nightmare to see what we have to repair.' At least they have a quiet summer to look forward to.

As Aintree winds down from the National, England's cricket groundsmen are well into their work – and worries – for the approaching season. No more so than at New Road, Worcester, where the county ground sometimes spends a good part of the winter under water. In the six months from 1 October 2000 the grass was invisible for 45 days and too waterlogged to walk on for a further 100 or so. Yet hard work, and a few days of drying wind, made it playable for the first match of the season in late April. Or it would have been if rain had not prevented play! Cricket curators have an almost impossible task. They have to produce a pitch on something resembling concrete, shave the grass literally back to ground level, and not use any water on it till the game is over. How would Alan Titchmarsh cope with that?

Meanwhile, to quote Mick Moore, long-serving curator at Oxford United's old Manor Road Ground, 'Every football groundsman in the country cannot wait for the season to end.' Not that this means a holiday. The players may shoot off on vacation, but the groundsman has little chance of relaxation. After struggling through the winter, the inter-season break is the time for renovation, for re-laying, for making things green again. Although the gap between the end of one football season and the start of the next seems to be getting shorter, at least there is no one playing at Highbury or Gigg Lane while restoration takes place. No such luck for the golf greenkeeper. The attitude of most golf club members is simple – we pay an annual sub and that means we can play all year round. Without a sympathetic committee, 'ground under repair' signs are the greenkeeper's only allies.

Summer arrives and out come the rain covers. Since 1981 county cricket regulations have enforced the covering of the centre square both before the match and during it when rain intervenes. Otherwise, if the hard, fast and true surface that the groundsman has laboriously produced gets wet, it can result in a real test of defensive batting against a turning ball on a spiteful, drying pitch. But such entertainment is no longer wanted and the 'sticky wicket' is a phrase that has disappeared from cricket's lexicon. Covers are less readily used on the outfield. Getting them on and off is seen as too time-consuming. Instead, reliance is placed on a variety of suckers, blowers, squeegees and supermops with occasional recourse to the old mop, bucket and sawdust. Tennis and bowls also require shelter from the storm, but here there has been a reluctance to embrace labour-saving technology. Dragging tarpaulins and their modern plastic equivalents across the courts and rinks remains the order of the day.

But summer can bring sunshine, sometimes too much if no rain has fallen. The task of the groundsman now is to get water onto the playing surface. This is easy enough for bowls rinks and tennis courts, not too difficult for cricket grounds, but a huge task for racecourses and golf courses where kilometres of pipes are required if the terrain is to be kept green. The irrigation system recently installed to service the six golf courses at St Andrews has 120 kilometres of pipework, 256 kilometres of cables and 4,000 pop-up sprinklers. It cost £2.5 million. Unless racecourses and golf links have an on-course water supply, there is also the cost of the water to be considered. Ludlow racecourse is fortunate in having a reservoir that holds nearly 16 million litres, the equivalent of about 160 mm of rain. For once golf clubs welcome bores; though outside the clubhouse. No wonder that in drought conditions, when the turf is as stressed as the greenkeeper, many local golf clubs concentrate solely on watering their greens to save the putting surface. Out on the course, they ban trolleys that might damage the fairways and allow preferred lies that don't.

All too soon – as some would argue – football is back and, as the autumn and winter rains begin, the groundsman's main concern turns to getting excess water off his territory. This is not surprising, given that a thunderstorm can douse a sports field with over 250,000 litres of water, most of it unwanted. Drainage is vital if winter sport is to proceed. Wet surfaces quickly churn to mud and can become unplayable. Although snow and ice catch the headlines, waterlogging is the most common cause of winter cancellations. In 2000, many minor leagues opted to declare all unplayed matches as draws so that the season could be completed. Without effective drainage, the ground can also become compacted as the players' boots compress the earth – just look at any goal-mouth in a well-used parklands football pitch – and this, in turn, renders drainage even more difficult. Horses weigh a lot more than footballers and compaction is a serious problem for racecourses, especially in winter. It used to be dealt with by hand forking, but aeration is now highly mechanised both on racecourses and football pitches. However, when the ground is very wet and densely packed, driving heavy equipment on the playing surface could actually aggravate the situation. Hence, when Murrayfield was flooded to almost a metre only a couple of days before the 2000 Rugby League Challenge Cup final, once the field had been gravity drained, the groundstaff saved the day by resorting to old-fashioned digging forks to ease the compression caused by the sheer weight of the water.

THE REFEREE'S DECISION IS FINAL

One game played on a wet pitch in winter can ruin it for the rest of the season. Once studs tear through the surface, there is virtually no chance of recovery until the growing season starts again. A scale of surface wetness traditionally used in agriculture and adopted by groundsmen runs as follows: hard and cracked, firm and dry but not cracked, firm and moist, moist and soft, squelchy in patches, squelchy all over, pools of standing water, and surface awash. (For horse-racing there are seven categories ranging from hard to heavy with firm, good to firm, good, good to soft and soft in between). To save their turf, groundsmen would prefer all matches to be cancelled when at least a third of the surface is squelchy and to have no more than one game a week when it is moist and soft. However, referees are the only people officially sanctioned to cancel a game and the only factor they take into consideration is the safety of the players. They cannot think about what might happen to the pitch in the future. So, at the very time that farmers would be herding their livestock off the soil, groundsmen have to allow footballers to pound all over it – a *prima facie* case of field abuse!

Neither old labour-intensive methods nor new spiking technology can be employed much on tennis courts and bowling greens during the playing season. Even if unlikely to be played on in heavy rain, tennis courts and bowling greens still need drainage to get the water away and prevent compaction when the players return.

The rinks and courts must also be flat, unlike football and cricket grounds which can utilise camber to assist in water removal. Wembley had a 28 cm fall from the centre to the touch-lines. This was nothing compared to Bolton's old home at Burden Park. Here, the slope was so steep that spectators at one side of the ground could not appreciate the dribbling skills of wingers on the opposite side, as only their upper bodies were visible.

Long grass of around 7 cm can act as an insulator against frost, but at that length is really suitable only for rugby pitches. Footballers and hockey players require a smoother surface on which to perform. In the good old days, layers of straw and the odd tarpaulin served to prevent football pitches becoming ice rinks, but such a method is ruled out now because of storage problems, labour costs and the risk of fire. In any event, at the élite level today's pampered professionals demand near perfection on which to display their skills and lessen the risk of injury. Chemical thawing agents got rid of the ice but damaged the turf and their use was soon abandoned. Insulation blankets of many forms have also been used, but generally handling difficulties, problems in keeping lightweight materials secure in the wind, and storage issues for pitch-size covers have limited their use to goal areas and particular trouble spots. More recently mechanised pitch protectors have been imported from Canada – where they really know about ice and frost – which raise the ground temperature a few degrees. Attempts to heat the pitch from above have proved wasteful of energy and often led only to barbecued turf! Heating from below has been more successful, first with electrical soil warming and later with pipes containing either warm air or water.

CLUBS 0 FROST 5

In the icebound winter of 1963 several football clubs used their ingenuity to try and beat the freeze. Blackpool tried flame throwers at Bloomfield Road, Chelsea a tar burner at Stamford Bridge, Leicester a hot air tent at Filbert Street, Birmingham a Danish snow-shifting tractor at St Andrews, whereas Wrexham relied on scattering 80 tons of sand on its Racecourse Ground. None of them worked.

Golfers, now clad in mass-produced, affordable waterproof clothing, still insist in turning out to play come frost or rain. The problem for the greenkeeper is that grass doesn't grow much in winter. Intensive use of the course at that time of year makes for worn patches on tees and greens, and divots that don't repair themselves even when replaced. Supplying players with portable synthetic mats, from which to play their iron shots, can ease the potential for corrugated fairways; worn areas on the green can be reduced by judicious manipulation of pin positions and the provision of temporary greens; and tees can be moved backwards and forwards or be replaced by driving mats. How the greenkeeper must long for snow. This is the only weather that can keep the golfers at bay, for not even coloured balls can be easily found in a snow-filled bunker!

Skirmishes can be won in the never-ending battle against the elements, but groundsmen cannot rest on their laurels – or spades. There is no room for complacency. Solving one problem often produces another. The 'theatre of dreams' must sometimes seem more like a stage for nightmares to those trying to preserve the pitch at Old Trafford. The drainage system there is good, so good in fact that it takes too much moisture away from the turf along with many of the nutrients. United's undersoil heating is also highly effective. It enables games to be played but dries the pitch excessively. Between them the drains and the heating have produced a dry and arid environment alien to the growth of grass. No wonder the pitch has had to be relaid so many times. Across the road, at the Old Trafford cricket ground, the use of a large balloon to cover the square in 1992 successfully kept the Manchester rain out, but it produced condensation that still dampened the turf. Elsewhere the laying of electrical cables or water pipes for underground heating has helped combat frost, but often ruled out the use of subsoil spiking to offset compaction. Generally, the more pitches, courses, greens and courts are made playable, the more they will be played on, leading to increased wear and tear. The groundstaff cannot win.

Groundsmen cannot stand still. Old methods and technologies do not always cope with new problems, such as those produced by the massive redevelopment of football and other stadiums in the past decade. Stadium design has outpaced turf technology. In 1998 Manchester United had two-thirds of its Old Trafford pitch in the shadow of the North Stand. Thus, in winter, most of the ground was frozen while the rest remained frost-free. Worse was to follow with the addition of two tiers on either end of the main stands. As with other clubs, such as Glasgow Rangers at Ibrox, United discovered that the high-tiered, wraparound stadium not only made winters darker at pitch level, but also detrimentally affected wind movement so that natural drying was reduced. This is no good for turf which needs light and air to grow and survive. In the reduced light regimes of the new stadiums there is reduced root growth, less dense coverage and greater susceptibility to disease. Ipswich eased the problem at Portman Road by installing transparent sheeting over the gable ends and stand roofing. Other clubs have used floodlights to lengthen the day. For most football groundsmen, however, the light has not yet reached the end of the tunnel.

SPORTS TURF RESEARCH INSTITUTE

In the 1920s two Yorkshiremen, Percy Clough and Norman Hackett, persuaded the Royal & Ancient Golf Club and the British Golf Unions that research was an essential prerequisite to the upgrading of golf courses. This led, in 1929, to the establishment of the Board of Greenkeeping Research at Bingley in West Yorkshire. By 1937 the staff had expanded from 5 to 23 and 395 golf clubs were visited and advised. In 1951, in recognition of the spread of research interests to other sports, especially football, the Board was reconstituted into the Sports Turf Research Institute. Current work, always aimed at practical application, concentrates on the turf needs of individual sports, grass cultivar testing, and commercial research into fungicides, pesticides and fertilisers. Results are disseminated through the Advisory and Consultancy Services division as well as through the Institute's annual *Research Journal* and quarterly *International Turfgrass Bulletin*.

Another problem for the groundstaff stems from the multi-use of facilities. Commercial considerations have tempted club executives to ground share, lease out to other sports for special events, and hold pop concerts and other extravaganzas. Four weeks of hospitality tents on Cardiff Arms Park during the Rugby World Cup killed off the playing surface there. The pitch had already suffered from having over 10,000 tons of constructional material stored on it while the Millennium Stadium was being built next door. Ground sharing, while bringing financial economies, can lead to excessive wear on a pitch. At one time the JCB Stadium, which co-hosts Wigan's football and rugby league teams, held 160 matches in just 18 months without a close season for renovation. Playing rugby league finals and internationals at Old Trafford football ground has caused severe damage to the grass because of the amount of side-stepping in the game. This produces divots and also causes root breakdown. On a normal ground, such as that at Wigan, where light and air can assist renovation and recovery, this might not be too much of a problem, but the shade and lack of wind movement in Manchester United's redeveloped stadium seriously hampers recuperation. Such multi-use of facilities is not a new phenomenon. In the late nineteenth century, Surrey's Oval hosted soccer, tennis and rugby as well as cricket, but pitch standards then were not as now. One cannot imagine Sir Alex allowing a flock of sheep to control the summer grass at Old Trafford!

Some environments have become so unfriendly for natural grass surfaces that their future is questionable. A few clubs are pioneering pitches where the natural grass is interwoven with plastic to ensure that chunks of it will never be dislodged, as they would on a traditionally sown or laid pitch. In early 2002, European giants Real Madrid and Juventus both came to Oxford United, not to play the Third Division strugglers, but to check on Kassam Park, Oxford's new ground and one of the best playing surfaces in Europe. It has been developed as a show pitch by a Dutch firm to demonstrate their new technology. If this innovative method of pitch preparation does not work in the new stadiums, and if artificial turf surfaces remain banned by the football authorities, then grass may well become simply one more consumable in the profit and loss account, an item to be replaced several times a season. At between £200,000 and £400,000 a time, this is relatively small beer to clubs spending millions on players and stadium development. Alternatively, as is happening already in some countries, the grass will be grown in trays – fairly large ones – outside the stadium, brought in for use on match days, and then wheeled out again for recuperation.

The first rule of greenkeeping is that there is nothing that can be done to change the weather. The task of the groundsman is to get on with the job of keeping surfaces playable no matter what the climate brings. From mops, buckets, and improved grass seed to sophisticated turf management, groundsmen have increasingly learned to apply science and technology to unending battles against the elements and the damage they cause, be it cracked pitches, patchy grass or waterlogging. Yet, as we shall see in the next chapter, for some sports there is another way of combating the climate's effect on turf – give up the grass and adopt artificial surfaces.

16

When All Else Fails

The obvious way to avoid the worst of the weather is to move indoors. Nineteenth-century pioneers did this with ice rinks for skating and curling, and swimming pools for aquatic pursuits. Playing indoors has now become the norm for these activities. Athletics tried the odd meeting in gas-lit halls but not till 1935 did the Amateur Athletics Association hold indoor championships – on the (covered-in!) Empire Pool at Wembley. Later they moved to a disused RAF hangar but, with the coming of transportable synthetic surfaces, city venues have been developed for high profile events. Such surfaces have also allowed other sports to seek shelter from the climate. Bowls went inside, first on jute mats but later on a variety of carpet-like surfaces, though the move has been limited to flat rink bowling; there are too many constructional difficulties in creating domed indoor crown greens. Tennis has also grasped the opportunity to free itself from swirling wind and dampening rain, though it began as an indoor sport, as real or royal tennis, on stone or wooden floors in the sixteenth century. Cycling was rarely an indoor sport in Britain – unlike on the continent – until 1994 when a purpose-built, indoor velodrome was opened in Manchester.

Not all sports can – or want to – move under cover but, as we have seen, staying outside on grass means gambling on the weather with the climate often the odds-on favourite. The modern alternative is to opt for a synthetic, weatherproof playing area. Salesmen of such surfaces often declare them to be better than the real thing. Groundsmen, fed up with the challenge of keeping a perfect sward under highly imperfect conditions, are likely to agree. Such surfaces need less maintenance, can stand high usage rates, are highly porous, and – an unfortunate sign of the times – can easily be repaired when vandals strike. Only the purists and the accountants complain; the former, because the lawn has been removed from lawn tennis and lawn bowls, the latter when the bill for renewal comes in, roughly every five years. (Though the bottom-line men would do well to remember that synthetics offer cost effective, all year round playable surfaces – and hence more paying customers – for the vast majority of sports.)

MAGIC CARPET

Artificial grass is a fibred carpet, generally of tufted construction rather than knitted or woven, made of polyamide (conventionally called nylon) or polypropylene. The carpets vary according to the sport to be played. Those for cricket and bowls have very short pile and are densely stitched. Those for soccer have a wide stitching to allow for the use of studded boots and have long pile strands of around 50 mm. Shockpads are often installed between the carpet surface and its base layers to make it more shock absorbent and hence more comfortable on the feet. Such pads are not normally used for tennis courts as they lessen the bounce of the ball. Although easier to keep in order than natural grass, the artificial version is not maintenance free and requires brushing, weeding and surface cleaning of moss and other impediments.

At a high-profile level, sports such as athletics adopted the new surfaces enthusiastically, well before the recent technological revolution. Most early athletic tracks were laid out on grass and were thus susceptible to wear and tear as well as drought and downpour. They were replaced by cinder tracks, but these also deteriorated during freezing or wet weather. Athletics therefore took advantage of the first generation of synthetic materials developed in the United States in the 1950s. So too did skiing, though for recreational and training purposes rather than competition. Hockey, often played in winter on the outfield of cricket grounds, was also quick to appreciate the new surfaces. Even though the early pitches were hard and very fast, bearing little resemblance to a grass playing surface, they still allowed games to be played in winter and in wet weather. Ironically it was water that turned out to be the key element in revolutionising hockey pitches. When water-based systems were developed to replace the older sand-filled ones, they were seized upon avidly and have now become the standard for all international matches. On these 'wet-field' surfaces, a 3 mm film of water is used to remove the danger of friction burns without hindering ball control. The main problem is that onfield sprinklers cannot be used to replenish the water at half-time. Rain guns have been developed for that task. Consequently, until warning systems were installed, many an innocent spectator got soaked while munching his or her pie. Hockey authorities now insist that all top rank pitches must be synthetic. In Britain, however, the relative cost of water-based pitches has restricted their number and sand-based ones are still dominant. Nevertheless, throughout Britain, hockey can now take place in most weathers.

Whereas both hockey and athletics have fully committed themselves to artificial surfaces, several other sports have accepted them whilst retaining grass as their preferred option. Championship bowls can be played on artificial greens, both flat rink and crown green, but most major outdoor events remain on grass. Within horse-racing, only 3 of the nation's 59 courses are the euphemistically termed 'all-weather' tracks. Originally introduced to allow winter sport when other courses were waterlogged or frost bound, they now offer all year flat racing, but this is generally regarded as inferior to that on grass, akin more to speedway than horse-racing. Winter races at these tracks no longer count towards jockeys' or trainers' championships and the Jockey Club has not yet allocated a Group One race to an artificial track. Tennis, on the other hand, has had hard court championships for decades, even while grass remained pre-eminent. Now, except at Queen's Club and Wimbledon, turf is no longer *de rigueur* for élite tennis. Even home Davis Cup ties are played indoors on artificial surfaces. So much for those critics who allege that the Lawn Tennis Association is caught in a time warp (apart from the name, of course).

For some sports wholesale adoption of synthetic surfaces is impracticable. They would be too hazardous for body contact games like rugby – unless the players padded themselves up as for American football – but are used for running and passing practice. In golf, variations in pin positions and the creation of contours on the putting surfaces are best left to grass surfaces. However, synthetics have been installed in practice areas where even one day's intensive use would be sufficient to damage turf for several months. In contrast, the artificial surface merely needs topping up with sand and is then ready to be hacked again. Where clubs have long-grass tees, they can be partly replaced with artificial turf for general use, leaving the grass in pristine condition for tournaments and other special events.

Others could use artificial surfaces but remain unconvinced, preferring tradition to technology. In 1971, Terry Venables wrote *They Used to Play on Grass*, a futuristic novel about football. A decade later, Queens Park Rangers, the team he was then managing, kicked off their 1981–82 season at Loftus Road on Omniturf, a sand-filled surface topped with polypropylene. A handful followed where QPR led the way. Luton, their opponents in the first league match to be played on an artificial surface, opted for a non-grass pitch when they refurbished Kenilworth Road in 1985. Next year both Preston North End and Oldham went for plastic, the latter because their pitch had always been prone to freezing. Finally, in 1987, Stirling Albion became the first and last league club north of the border to adopt an artificial surface. Although these pitches enabled matches to go ahead when they were called off elsewhere, there were complaints of the ball bouncing too high and running too quickly as well as 'plastic burn' injuries incurred in sliding tackles. Nor were goalkeepers too happy with diving around on plastic. Both the FA and its Scottish counterpart eventually decided that artificial surfaces were not to be, at least for professional soccer. Stirling's three-year trial was not extended and in England a phasing out of the synthetic surfaces began from 1991. The last match on artificial turf was played on 18 May 1994 when Preston met Torquay in a Third Division play-off. Top-level pitches must now conform to FIFA rules and, although that body has agreed to some experimentation with the new generation of artificial surfaces, currently grass is the only option for élite matches. However, many clubs are installing artificial grass with rubber infill for light training to relieve pressure on grass pitches when they are frozen or wet. At least football put its toe in the water before pulling back. Cricket at the highest level has never even reached the beach. The traditionalists argue that the hard, true bounce of a pacey turf wicket cannot be replicated on any other surface. Bear in mind, however, that they also said that one-day cricket would never catch on!

It is at the lower levels of sport that artificial outdoor surfaces have been widely adopted. For local authorities and those schools that have not sold off their playing fields, the synthetic pitch is a valuable labour-saving device: not that they are maintenance free. They have to be top-dressed with sand, brushed to work it into the carpet pile and checked for tears, especially along the seams, but this is significantly less work than trying to keep a turf pitch playable in all weathers. Unhappily, some cannot afford even that level of maintenance after the initial purchase and there has been a rise in injuries, particularly to the knees, when carpet pile has collapsed and reduced the grip offered by the artificial surface. The second generation of synthetic surfaces has generally moved away from a 'one product serves all' approach and manufacturers have increasingly tailored their artificial turf to the specific needs of individual sports. Nevertheless, sand-filled carpet pitches are capable of multi-sport use where economics has priority over exclusivity. An artificial soccer pitch, for example, needs between 1,200 and 1,500 hours annual use to be cost effective; if it can also be marked out for hockey, netball, tennis and children's sports, all the better for the budget.

At the top level of football, the quagmires of the 1960s and 1970s are long gone. Plastic had offered one solution, but salvation has come with improved pitch construction and drainage, the breeding of new grass strains, the development of maintenance equipment, better fertilisers and pesticides, and enhanced training of groundstaff. Yet all this costs money, expenditure way beyond the budgets of most local councils and state schools. For them, artificial surfaces are the rule. A generation is growing up rarely having played organised sport on grass. Yet, unlike their predecessors, they can turn out in all weathers – unless they brought a note from Mum!

HAVING IT BOTH WAYS

Built initially to host the 1999 Rugby World Cup and assisted by £46 million of Lottery funding, the Millennium Stadium, built in Cardiff alongside the River Taff, is Britain's first – and so far only – sporting arena with a sliding roof designed to keep out the elements. Melbourne in Australia already has two of its own. With the roof open, the grass playing surface can gain the benefits of light and fresh air but, to be doubly sure, the grass is in moveable modules that can be rotated from shadow to sun. A sprinkler system and artificial air movement also help keep the arena grass in good shape. And, in a classic case of sporting belt and braces, 1,000 square metres of turf are tended in pallets offsite. With the 105-metre by 80-metre roof closed – winching takes 20 minutes – not only can rugby and football be played without the weather adversely affecting skills but all kinds of indoor events can be hosted. Sounds good. A financial dispute with the contractors, however, hindered the correct rotation of the grass modules which led to pitch problems appearing towards the end of the rugby World Cup. Then the roof jammed open in the weeks before the FA Cup final (transferred to Cardiff while a decision is awaited on the redevelopment of Wembley) which allowed several Welsh downpours to drench the turf and necessitated it being replaced.

THERE MAY BE TROUBLE AHEAD

17

True Grit: The British Spectator

Athletes and groundsmen are not alone in their battles with the elements; spectators also suffer when the weather is unkind. Getting to events, parking at sports venues and keeping dry and comfortable when there are just some of the difficulties faced by the travelling supporter. Many spectators cannot even rely on the luxury of a grandstand or covered area from which to view their sport. Although the lucky ones are housed in modern stadiums, there are still thousands of fans on open terraces and unroofed stands. Even state-of-the-art football grounds are not weatherproof and anyone who has sat in the front rows of the average grandstand knows how exposed they are on wet and windy days. Some will never find shelter because the event takes place beyond the confines of a pitch. It is no coincidence that large multicoloured umbrellas were popularised on golf courses and green wellies at horse trials, point-to-points and other country pursuits!

It was definitely welly boot weather in April 2000 for the British Grand Prix at Silverstone, one of the worst weather-related fiascos of recent years. Exceptionally wet conditions and a high water table swamped 200 acres of fields normally used for car parking, forcing organisers to ban cars for the practice and qualifying day. Up to 30,000 vehicles had been expected, but only coaches and bicycles were allowed through the gates while efforts were made to lay tonnes of straw and hardcore material. Additional parking for race day was arranged at nearby Towcester racecourse and a local aerodrome, but many motor racing fans with tickets for the practice session had to be satisfied with refunds. Formula One drivers had to cope with pools of water and slippery conditions on the track; David Coulthard had to be rescued by a vehicle with tractor-like wheels when his car strayed onto the sodden grass around the circuit during practice (he later went on to win the event).

For spectators, worse was to follow. On the day of the race, 25-kilometre traffic jams built up on roads leading to the Northamptonshire village as ticket holders set off early to get there. Fog was an additional hazard, although most problems were caused by impossibly slow manoeuvring and parking in the wet, muddy conditions. Many abandoned their cars and trudged the last few miles, arriving long after the race had started. At least they made it. By noon, police had set up an 8-kilometre exclusion zone around Silverstone and advised drivers to go home and watch the race on television as they had no chance of getting to the track. Fans, who had paid up to £200 for a ticket, were naturally upset. The chaotic scenes were blamed on a number of factors but focused on the very wet spring and the ill-advised transfer of the race from July to April. It reverted to its traditional slot in the Grand Prix calendar the following year but as all sports fans are aware, the British summer does not guarantee rain-free weather. July 2001 saw more rain during the preliminary days but thousands of square metres of reinforced mesh in the car parks – at a cost of £750,000 – and the introduction of both a one-way traffic system and a park-and-ride scheme ensured that the PR disaster of the previous year was not repeated.

Sadly, the Silverstone débâcle is not an isolated incident. Dreadful weather in April 1986 put the car parks at the Badminton Horse Trials beyond use and would-be spectators had to walk from the nearby village. Tennis fans were advised to use public transport during the first week of the 1987 Wimbledon championships because heavy rain had turned parking facilities into mud baths. A match between Northampton and London Irish rugby clubs was postponed in April 1998 when the access route to the car park became impassable in torrential rain, while July 2001 saw flooded roads and parking areas before the start of the Scottish Open at Loch Lomond; spectators were warned not to travel until mopping up operations had been completed. Episodes such as these resulted in inconvenience for the paying public and sometimes led to smaller crowds and financial losses for promoters, all because the weather turned nasty and saturated ground facilities. Given the unpredictable nature of the British climate, it seems surprising that so little effort appears to be spent on ensuring that grassy car parks remain viable in wet weather. Alan Lee, in his recent guide to British racecourses, tells the hopefully apocryphal story of tractors towing cars *into* as well as out of the main car park at one venue – but this cannot be assumed! His assessment of car parking facilities gave 5 marks or less out of a possible 10 at a fifth of the 59 British racecourses – comments ranged from 'can be a bog', through 'pitted and puddled' to 'a field, often waterlogged'. It is astonishing that spectators can be prevented from attending sports events in the twenty-first century by nothing more drastic than wet weather.

Moreover, where would sports organisations be without them? Every winter, the call goes out to football fans to help clear snow-covered pitches and they are usually rewarded with a free ticket to the game. The derby match between Birmingham and West Bromwich Albion in December 1993 went ahead after a local radio appeal brought 300 supporters to the St Andrews ground to help remove snow from the pitch. At the end of January 1996, snow ploughs and volunteers spent a whole day clearing a 15 cm fall from the Bramall Lane pitch and seated areas to enable a televised FA Cup tie between Sheffield United and Aston Villa to go ahead. Even grounds with protective covering can be in trouble without extra manpower. West Brom relied on fans turning up at The Hawthorns with brushes and shovels to remove snow from the pitch covers in 1997. Sometimes willing spectators can actually help their team to victory. When a thunderstorm at The Oval turned the playing surface into a lake in August 1968, at least 50 volunteers from the crowd assisted groundstaff in their efforts to mop up as quickly as possible. Speed was essential. It was the final Test between England and Australia and the home side, trailing 1–0 in the series, was in a potentially winning position at lunch on the last day, with Australia on 82 for 5, needing a total of 352. Fortunately the run-ups had been covered, but it still took staff and assistants all afternoon to restore the wicket and outfield. Play finally resumed at 4.45 p.m. and in a hectic hour of play, England took 5 wickets in 20 overs, thanks largely to Derek Underwood, and squared the series with 5 minutes to spare.

Where weather and cricket are concerned, however, things are more likely to end in tears than triumph. The national summer game can be very frustrating for all concerned, not least the spectator. The joys of an average British season ensure that many hours will be spent huddled under umbrellas or peering gloomily from the shelter of the grandstand while the rain alternately cascades and drizzles down and the dark clouds refuse to budge. Hopes rise as the umpires, captains and sometimes entire teams appear . . . and disappear. Occasionally it is all too much for the disgruntled fan. In the final Test between England and Australia at The Oval in 1921, play was interrupted by a heavy shower on the second day of a three-day match. When the captains inspected the wicket, the Australian Warwick Armstrong is said to have gestured as though wringing water from his hands while England's Lionel Tennyson made no response. The crowd wrongly interpreted this scene, assuming that Tennyson judged the wicket playable while Armstrong was indulging in gamesmanship. They invaded the pitch and gathered in front of the pavilion, shouting and booing. Play restarted at 5.15 p.m. but, although 471 runs were scored on the final day, the match was drawn; not that this mattered much as the series had already been lost 3–0.

Frustration at the weather boiled over more seriously during the Centenary Test between England and Australia at Lord's in August 1980. The match was ruined by rain, with ten hours lost, and ended in the inevitable draw but it was also marred by spectator violence. In a most un-English and unedifying episode, umpire David Constant was assaulted outside the Long Room by several angry MCC members as he returned from a fifth pitch inspection on the Saturday afternoon. The incident was recounted at length by the other umpire that day, Dickie Bird; he devoted an entire chapter of his autobiography to what he called the worst day of his cricketing life. What actually happened is debatable and depends on whether you believe Bird's account, that of England captain, Ian Botham, or any one of several newspaper correspondents. A majority wrote that the umpires were given a police escort for the remainder of the day. The match was also noteworthy for other reasons, marking 7,000 runs in Test cricket for England batsman, Geoffrey

Boycott, and the final Test match commentary by the legendary John Arlott. Pity about the weather!

Less violence but similar frustration was shown during the second Test against the West Indies in the poor summer of 1988. Once more Lord's suffered from rain and bad light and the crowd, fed up with continual stoppages, showered the pitch with seat cushions in scenes reminiscent of Wimbledon 50 years earlier. On that occasion the normally genteel tennis crowd had hurled cushions onto No. 1 Court after rain had interrupted a Davis Cup tie. More recently, a day of rain on Centre Court in 1985 produced a barrage of slow hand-clapping and only 18 minutes of tennis, followed by cries of derision as first John McEnroe and then Alan Mills, the referee, decided that the grass was too wet for play.

Followers of winter games such as football and rugby can be just as dissatisfied with the weather and its consequences. These range from games rendered farcical by wet, icy or snowy grounds to matches called off at the last minute because of late pitch inspections or a sudden deterioration in conditions. Even more frustrating can be postponed games, rearranged for midweek winter evenings, when common sense suggests that they could easily be off again. If long-distance travel is involved, fans have the added inconvenience of cold, dark journeys. It can be difficult enough on a Saturday, as Sheffield Wednesday supporters discovered in December 1990. Twelve coachloads set off from Yorkshire for the game against Bristol City but never made it through the snow. It is not surprising that supporters become equally incensed at games being called off when they might have gone ahead and others taking place when they should have been postponed.

While there are no easy answers there are always many opinions. In the harsh January of 1987 the manager of Norwich was quoted as saying that it was unfair to expect fans to travel to games to watch 'something which is hardly a game at all because of the treacherous conditions'. In the same week Hearts, from the Scottish First Division, were upset that their home game was called off more than 24 hours before the fixture; they thought it was too early to pass judgement on the pitch. By January 1997 the secretary of Crystal Palace stated that postponements could no longer be left to the last minute because fans threatened to sue for travelling expenses, which is precisely what the Montrose supporters club intended to do in January 1998. When their Scottish Cup tie against Clydebank was postponed 25 minutes before kick-off, they were furious at having made a 400-kilometre round trip for nothing and considered seeking compensation from the Scottish Football Association.

It is this type of late postponement which really upsets fans. When Oldham were scheduled to play Luton under floodlights in season 2000–01, the pitch at Boundary Park became so waterlogged after heavy rain that the match had to be called off despite the fact that the visitors and their supporters had already arrived at the ground. A postponed third-round FA Cup tie between Brentford and Manchester City in January 1997 was due to kick off at 7.30 p.m. but fell foul of the weather at 5.30 p.m. with many visiting fans already at the stadium. Although it had passed an inspection in mid-afternoon, one side of the pitch had become affected by frost following a sharp drop in temperature. Two Premiership Boxing Day games were called off at the last minute in 1995, with West Ham v Coventry postponed only 70 minutes before the scheduled kick-off. The Coventry manager, Ron Atkinson, was quoted as saying that the pitch was too hard for football and that there was no point in playing if the fans could not get value for money. Yet 130 kilometres away the game at Southampton went ahead on a terrible pitch. Decisions sometimes appear to be arbitrary and it is the referee, naturally, who bears the brunt of any criticism by press and public. Opinions vary even amongst officials. The referee called off a rugby match between Bristol and Northampton in December 2001 only an hour before kick-off as he considered that the pitch constituted a danger to players. It had already been examined at 12.30 p.m. by the Bristol coach and a former international referee, who had both passed it fit for play.

Very occasionally, the effort made by fans to turn up in bad weather is rewarded. In March 1952, England played Ireland at a snowbound Twickenham in a mixture of snow showers and intense cold. The game probably should not have gone ahead but the secretary of the RFU justified the decision to play on the grounds that over 4,000 Irish supporters had travelled specifically for the match. In the modern professional era it is unlikely that such risks to players' well-being would be taken. The power of the paying spectator may also have influenced the decisions to play a rugby match between New Zealand and Scotland in 1975 and to allow the Japanese Grand Prix to take place in 1976 (see Chapter 22).

Fans are stoical. They plough through snowdrifts or glutinous mud; they stand, unprotected from the elements, in howling gales, pelting rain or horizontal sleet; they sit in dismal, leaky grandstands – all for the sake of sport. This was particularly true before mass access to television. A record crowd turned out in March 1914 in appalling weather to see the Irish football team draw 1–1 with Scotland to win the Home International Championship for the first time. A post-Olympics athletics meeting at the White City in 1952 between the USA and the British Empire saw a capacity crowd of 45,000, in spite of torrential rain, with a further 20,000 unable to get in. Even in the twenty-first century, supporters will brave the elements to cheer on their team. When the European women golfers took on the Americans in the sixth contest for the Solheim Cup in October 2000, five-figure crowds trudged daily through rain and mud at the Loch Lomond course. There were 15,000 present on the Saturday of the tournament when play was halted at 3.30 p.m. because of the atrocious conditions. They even paid for the privilege of getting soaked for six hours, but were finally rewarded the following day with a European victory.

It is only within the past ten years that some sports events have deigned to offer compensation to long-suffering fans when the weather has washed out a day's play. There were ugly scenes outside Edgbaston in June 1992 when 15,000 Test ticket holders were told they would not be getting a refund because two balls had been bowled. The Test and County Cricket Board had taken nearly £300,000 in advance ticket sales for the second day of the match between England and Pakistan but, as three minutes of cricket had been played, they stood by their reimbursement clause which specified no play whatsoever. This policy was later altered to allow full refunds if there were fewer than 10 overs played and half the ticket value if between 10 and 25 overs were bowled. Sensitivity is now a watchword in cricket circles. When Surrey played Hampshire at The Oval in April 2000, the match was reduced to 10 overs because of a sodden outfield, although the day had been mostly sunny. Play finally started at 5 p.m. and finished in failing light but Surrey, aware that disgruntled spectators had been hanging around for six hours, offered to return admission money (while stressing that this did not set a precedent).

Other sports face similar dilemmas. Racing at Brighton was abandoned after one race in May 1994, heavy rain having turned the track into a mud-bath, but no refund was offered to racegoers. A bookmaker was quoted as saying that the course was unraceable at the outset but that management were determined to pocket the punters' money before declaring the meeting off. Abandonment after the first race still leads to loss of admission money at many of the smaller racecourses across the country, but larger courses such as Cheltenham, Newmarket and Goodwood now offer 50 per cent refunds if racing is called off before the third race.

And then, of course, there is Wimbledon. It was 1987 before the All England Club began to consider reimbursement for spectators when play was cancelled because of bad weather. The following year, although refunds were still unavailable, ticket holders on the show courts were guaranteed priority for next year: providing they paid again! Today full compensation is offered if there is less than one hour's play because of rain and 50 per cent if play is restricted to between one and two hours.

Sometimes watching on the box seems preferable to braving the weather, even for die-hard fans like those of the 'Old Firm'. A poor attendance for the Scottish Cup final of 1977 between Rangers and Celtic was blamed on a combination of a miserable day and live television coverage (Celtic won 1–0). The previous September had seen George Best sign for Fulham, but even his pulling power was insufficient to attract the big attendance hoped for by the home side. When an afternoon of torrential rain was followed by a cold, damp evening, less than 15,000 turned out to see Fulham draw 2–2 with Bolton, only half the anticipated crowd. Television coverage and wintry conditions in January 1987 were also held responsible for the pitifully small crowd (less than 18,000) that turned up at Highbury to watch Arsenal draw 0–0 with unfashionable Coventry. The home side, accustomed to attendances of double this figure, claimed around £45,000 (£75,000 today) from a fund set up by the Football League to compensate clubs whose gates suffered from the presence of television cameras. Considering the conditions under which many twentieth-century spectators have had to view their sport, it is surprising that more have not taken the soft option of watching from the comfort of the pub or the living room. Let's hear it for the British sports fan, the epitome of true grit!

18

Brain Beats Rain: The Duckworth/Lewis Method

'Y ou play for five days and still nobody wins!' Foreigners, Americans especially, are bemused by Test cricket. In the States there is no such thing as an honourable draw. Even tied games are not permitted in their major team tournaments and overtime is played to ensure a winner emerges from the fray. Yet in England the drawn game, often because of weather problems, remains a permanent feature of both Test and county cricket. Only in the one-day, limited overs game that has developed since the 1960s has a positive result been seen as imperative. Here a winner was guaranteed: provided that it did not rain.

If rain intervened, the solution was to ask the side batting last to achieve a run rate (in its reduced number of deliveries) marginally better than the average rate of scoring during the whole of the first side's innings. As teams usually paced themselves with higher scoring coming later in the innings, the balance too frequently lay with the side batting second, unless rain simply delayed the start of a match and both teams knew from the beginning that they each faced a reduced number of overs. Despite the absurdity of the situation, it took three decades for a fairer system to be found.

The Australians experimented with a 'discounted least productive overs system' but on occasions this actually aggravated the problem. A farcical example occurred during the 1992 World Cup match at the Sydney Cricket Ground between England and South Africa. When a heavy but brief shower halted play at 9.52 p.m. the South Africans required 22 runs from 13 balls. England knew that, under the rules governing the competition, if the match was called off they had won. If the players did return to the field the South Africans, with four wickets left, were confident that they could achieve the target. However, they had miscalculated. One clause of the match regulations deemed that when time was lost to rain the number of overs to be bowled should be reduced at the rate of 14 per hour. The rain stopped. One over was deducted from the South African's innings and England's least productive over – a maiden – was knocked off the target score. Now the batting side required 22 runs from 7 balls, a feasible task but infinitely more difficult than from 13 balls. When the arguments had subsided even more time had been lost. Another over was deducted; this time, one run was discounted; and it was announced to a disbelieving crowd that the South Africans would come out to face just one ball, from which they still had to score 21 runs! England won.

Two statisticians came to the rescue. Frank Duckworth, a freelance statistical consultant who had taken early retirement, and Tony Lewis, a lecturer in mathematics at the University of the West of England, used their computing skills to devise a more equitable method of determining the destiny of rain-affected matches. Statisticians have always had a role in the subculture of cricket – in working out averages and providing historical comparisons and numerical oddities. Now the number crunchers were actually to help determine the results of games! Their basic premise was that if a team knows it has only 20 overs to bat, it will go at it more furiously than if it had 40 overs. Their formulae, based on first innings' statistics from several hundred matches played from the 1990s onwards, made allowances for downpours and showers, taking into account when they occurred and the number of wickets that had fallen at the time. What many cricket observers could not grasp was that the logic of this led to sides being asked to make significantly more runs than their opponents from the same number of overs.

Brought into the English domestic game for the 1997 season, the new method of calculating the revised run rate was trialled on England's tour of Zimbabwe the preceding winter. New Zealand, hosting the tourists after their Southern Africa visit, were given the option of playing under Duckworth/Lewis, but, to their later chagrin, they turned the offer down. That was their first mistake. They won the toss and chose to bat. That was their second. The conditions were extremely humid and rain was expected. It came: as did an England victory. When rain first intervened, England had scored 47 runs from the 6 overs they had faced. On resumption, under the old-fashioned method, they were allowed 20 overs to score a further 85 runs, using a ball that was liable to get wet and hamper the bowling side. The target was achieved with 39 balls to spare. Duckworth/Lewis would have required 117 more runs to be obtained, a significantly harder task.

Although its introduction left many people, not just spectators but also players and umpires, mystified at some of the calculations, it has now become accepted, if not yet fully understood. In April 2001 the formulae were amended on the basis of the experience of some 1,500 matches. Of course, to cricket purists, brought up on the wearing of whites and the tactics of the first-class game, Duckworth/Lewis is simply another layer of artificiality in what they regard as an aberration in the first place.

19

Fantasy Football: The Pools Panel

A s the cuckoo heralds the coming of spring, so the calling together of the Pools Panel signals that winter is upon us. At least it did until other functions were added to its task of 'forecasting' the results of games postponed by the rigours of a British winter. On every Saturday last season, from the first in November to the last in April, a group of men met in a London hotel. Provided with information on form, and sandwiches for sustenance, they were locked away before kick-off and not released until after half-time in the latest match to be played that day. In the interests of security – with millions of pounds at stake – no one else was allowed into their room. Their task was to predict the results of matches on the pools coupon that were not actually played that afternoon.

The origins of the Panel lay in the savage winter of 1962–63 that wrought havoc with the football fixture lists. There had been a week of snowy weather in mid-November and a spell of severe frost in early December, hardly harbingers of what was to come. From Boxing Day until 2 March the ground remained white throughout many parts of the country, the longest unbroken period ever recorded. Overall, it was the coldest winter of the twentieth century and, in central England, it was second only to the winter of 1739–40 in over 300 years of recorded statistics. The snow and ice led to the longest ever third round of the FA Cup. It lasted 66 days, from 5 January to 11 March, involving some 261 postponements. In all that winter nearly 500 league and cup matches were disrupted. After three weeks of declaring the football coupons void as at least 30 matches were off, the pools companies decided to salvage their livelihood from the big freeze. A panel of experts was brought together to predict the results of the postponed league programme of 26 January.

There was immediate criticism. Anti-gambling moralists seized on the distinction between the element of skill required in betting and the absence of such a requirement in the purchase of a raffle ticket, arguing that the use of the Panel transformed the pools into an illegal lottery. Customs and Excise also wondered if they could collect the relevant duty under the Pools Betting Act if the forecasts of the Panel were deemed illegitimate. However, the legal advice given to the Pools Promoters Association proved correct in that, as the Panel was supposed to use its collective judgement in making the predictions, punters too could assess form and other variables in making their selections.

While some thought the whole set-up a work of fiction – fantasy football at its worst – others felt that the Panel did not go far enough. Some wanted it to give actual scores rather than just results. Cliff Lloyd, secretary of the Professional Footballers Association, thought it was a great pity that actual scores were not predicted, as this would have assisted the many supporters' clubs who operated 'pontoon pools' in which the winner was the first team to aggregate 11 goals.

Others felt that it would be more equitable if the Panel forecast the result of *any* postponed match that was on the Saturday coupon, not just the 30-plus cancellations that were necessary before the system came into operation. Not till the 1980s was this change brought in. It led to situations in which the full Panel met in a West End hotel to debate the likely outcome of just one match. Sometimes these were not even mainstream games. In March 1994 the Panel discussed the match between Knowsley and Matlock in the Northern Premier League and in December 1988 it had to decide on the probable result of Hyde versus Rhyl in the HFS Loans League, a fixture postponed because Rhyl had a Welsh Cup third-round replay.

Lord Brabazon of Tara, an ex-government minister who had celebrated his 70th birthday by bobsledding down the Cresta Run, chaired the first Panel. He had also been president of the English Golf Union, the Royal Institute and the Royal Aeronautical Society. Although he might have lacked football expertise, the other Panel members provided this in abundance. They were former referee Arthur Ellis, and former professional players, Tom Finney, Ted Drake, Tommy Lawton, all England internationals, and George Young, ex-Scotland captain. Ellis, who had hung up his whistle in July 1961, was a world-renowned referee who had officiated at the 1953 FA Cup final. Finney, from Preston, had gained 76 caps for England before his retirement in 1960 and was acknowledged as one of the finest international wingers of his generation. Drake, a former England centre-forward, had been associated with football for over 30 years, latterly as manager of Chelsea, whom he had left in September 1961. Lawton, also a centre-forward, had played for Burnley, Everton, Chelsea and Arsenal, gaining 23 international caps. The defence had only one representative, ex-Rangers captain George Young, who played at either full-back or centre-half for both his country and club, winning 52 caps.

The members of the Panel were each paid £100 (£1,250 today), though Lord Brabazon was reported to have received a higher fee, which he donated to charity. The Panel was also wined and dined in a top London hotel. The total cost was estimated by the press to be around £1,000 (£12,500), a small price to pay to protect an industry with gross annual receipts of £80 (£1,000) million and which brought the Post Office some £20 (£250) million in revenue. They forecast 23 home wins, 8 away wins, and – some would say – a convenient seven draws. For that first week many punters appear to have adopted a wait-and-see policy, but the results announced by the Panel provided dividends that were regarded by the Pools Promoters Association as 'very satisfactory considering the circumstances'.

With space to fill, several local newspapers were inspired to write tongue-in-cheek match reports of an imaginary game, including details of the goals, though the Panel itself only stated that the result was a home win, away win or draw. At the stroke of a pen, marginal victories were turned into massive triumphs providing, of course, that the Panel had given them the win in the first place. The apocryphal story which did the rounds was that, on hearing the Panel's decision, one player applied to his club manager for his win bonus, only to be told that he had been dropped for that particular game.

Initially the BBC was unsure whether or not even to broadcast the results of what *The Scotsman* newspaper, descending to the language of the tabloids, labelled 'the sham soccer pools'. However, from the beginning the commercial media companies, especially ITV, wanted to televise the deliberations of the Panel. The Pools Promoters Association, believing that this would place undue pressure on the Panels, resisted these overtures. In the event the first meeting was televised, but all viewers saw was the Panel's emergence from a soundproofed BBC studio.

Over time the Panel has been brought into play for reasons other than the weather, most dramatically for the first Saturday in September 1997, following the cancellation of all games as a tribute to Diana, Princess of Wales, killed the previous week in a car crash in Paris. More generally, the reshuffling of fixtures as television companies demand that games are played at times other than the traditional Saturday afternoon has forced the pools companies to extend the role of the Panel, so that the results of Sunday or even Monday matches can be determined in advance. Additionally, as the pools companies have diversified the old 'treble chance', they are now called upon to forecast half-time results and whether drawn matches had any goals in them.

Their most intensive weather-related day's work came on 27 January 1996 when all but three FA Cup fourth-round matches and one Endsleigh League game in England were postponed. Scotland fared a little better: five Tennents' Cup third-round games and one Second Division match were possible.

Although some eminent public figures have chaired the meetings – Lords Preston and Bath, for instance, as well as air ace Douglas Bader, trade union leader George Woodcock, and Sir Alan Herbert – the Panel has relied heavily on those with an involvement in the professional game. In the mid-1990s Arthur Ellis was still a member, a nice if somewhat irregular little earner, alongside 1966 World Cup winners Roger Hunt and Gordon Banks, ex-Scotland goalkeeper Ronnie Simpson and journeyman player Tony Green, who played for Albion Rovers, Blackpool and Newcastle.

The Panel is supposed to take into account such factors as the current home and away form of the respective teams; the results of previous meetings between the teams over the past five seasons; the effect of changes in team membership; known injuries; tactics and set-piece plays; and the disciplinary record of the allotted referee. Where opinion is divided, the majority prevails or (ultimately) the casting vote of the chairman. Occasionally the Panel's selections have bemused students of the game who felt that an illustrious group of football experts should not come up with some of the bizarre outcomes that ruin the formbook. One writer to the *Daily Telegraph* in February 1996 complained that some of the Panel's decisions for the first Saturday of that month defied all logic. Gillingham had won 10 out of 12 home games and conceded only 3 goals, yet the Panel forecast a score draw with Barnet, who had lost 3–1 at home the previous week and had only a modest away record. Then there was Hednesford at home to Stevenage. Hednesford's home record was W6 D2 L4 while Stevenage's away record was W7 D3 L1. How could this possibly be seen as a victory for the home side? North of the border, Hearts had beaten the mighty Rangers 3–0 at Ibrox the previous week, but the Panel gave struggling Partick Thistle a score draw on their visit to Tynecastle! Of course, pools winners never questioned the decisions.

Despite the amount of money hinging on its verdict, the Panel has never had to account publicly for its decisions. Generally, no details of its discussions are given, but in 1973 the group allowed a television documentary to be made of its deliberations for the postponed Newcastle United v Chelsea match. The predicted result was a home win, with a privately agreed result of either 2–1 or 3–1 (remember that the scores were never officially announced). The game was eventually played ten days later and, as part of the 'experiment', the Panel was reconvened prior to the match. This time, possibly influenced by tabloid publicity over an alleged extra-marital affair of a Magpie defender, they opted for a score draw. The actual result was 1–0 to Newcastle.

Following the initial forecasts of the Panel, the first of the phantom games to be subsequently played was Falkirk's 2–0 defeat of Celtic on 29 January 1963, two days after the Panel had forecast a draw. Ever since, the predictive ability of the Panel has been under scrutiny. Some statistical anoraks have attempted to quantify the accuracy of the Panel's predictions. In the 1981–82 season the Panel convened five times and 'invented' 182 results. When the matches were actually played, its success rate for home wins was 49 per cent, for away wins 46 per cent, for score draws 23 per cent, but they failed to predict any of the no-score draws. Their overall hit rate was around 39 per cent. They were also less accurate as they progressed down the leagues, with 43 per cent correct for the First Division, 41 per cent for Division Two, 35 per cent for Division Three and 20 per cent for Division Four. Another survey, this time of the 1984–85 season showed that the Panel was called on to predict the results of 161 games from the English and Scottish Football Leagues, 4 non-league fixtures and 3 FA Cup ties. Their success rate was 42 per cent in the English League matches, 62 per cent in the Scottish, and 58 per cent in the non-league games. They got all three cup games wrong. In fairness, it must be noted that often the games took place several weeks after the initial postponement, during which time players could have lost form or been transferred, teams restructured and, increasingly in modern football, the manager sacked.

For the record, here are the forecasts that the Panel made for the third round of the FA Cup on an icy first Saturday in January 2002 and the actual results when the matches were eventually played.

	HALF-TIME	FULL-TIME
Brighton v Preston	score draw (0–1)	score draw (0–2)
Cheltenham v Oldham	draw (1–1)	away (2–1)
Southend v Tranmere	away (0–0)	away (1–3)
Coventry v Tottenham	home (0–1)	score draw (0–2)
Crewe v Sheffield Wednesday	draw (1–1)	home (2–1)
Darlington v Peterborough	home (1–0)	home (2–2)
Rotherham v Southampton	draw (1–0)	draw (2–1)
Stockport v Bolton	away (0–2)	away (1–4)
Walsall v Bradford	away (0–0)	home (2–0)
Wimbledon v Middlesbrough	draw (0–0)	home (0–0)
Wycombe v Fulham	draw (0–0)	draw (2–2)

In the light of these results, the arguments of the opponents to the introduction of the Panel in 1963 seem reasonable. Out of the 11 half-time predictions, the Panel got four correct and a similar number for the final results. In only one game, Stockport v Bolton, did they successfully forecast the correct result at both half-time and full-time. Clearly, in asking punters to second-guess a Panel that apparently has little predictive ability, the pools promoters have turned the coupons into lottery tickets.

20

Covering the Cost: Weather Insurance

International cricketer Len Braund loyally served Somerset, his county cricket employer, for over a decade and was rewarded with a benefit match during the 1904 season. From this he would retain the net profits, a welcome nest egg for the future in the days before old age pensions and the welfare state. Some players did very well from such matches, most notably Yorkshire's George Hirst, who took home £3,703 (£255,600 today), the highest total before the First World War. Hirst's benefit was in 1904, the same year as Braund's, but the Yorkshireman's match against Lancashire was blessed with fine August weather, unlike poor Len's. It rained on his parade at Taunton and no play was possible. Legend has it that Braund even took blankets from his hotel in an abortive effort to dry the pitch. The match was abandoned. At least the teams obtained some championship points for a drawn game, but Braund was actually out of pocket as he still had to pay the wages of his team mates, the umpires' match fees, and the hire of the ground. He was not alone in his misfortune. Ten years previously, four of the eight benefit matches arranged for the 1894 season were partly disrupted by rain and showers. No wonder the apocryphal dressing-room story developed of players being asked if they wanted a second benefit and replying that they could not afford it!

Inclement weather is less of a problem for the beneficiary today as his fund is built up over the year from dinners, collections and auctions and is thus less reliant on the proceeds of a single game. In any case, a complete wash-out or a serious disruption can now be covered by insurance. These days it is not only cricketers who take out such protection, and not just rain that is covered by the policies. Car rallies are frequently insured against winter weather as although some snow and ice can add excitement to an event, too much can sometimes force its cancellation. The routes used are often in mountainous areas susceptible to icy conditions, along tracks that would not be gritted, and they are often held at weekends when less effort is made to clear the roads. Rugby matches are more vulnerable to cancellation than soccer games as a frozen ground is simply too dangerous to play on when bodies are being hurled around. Hence many professional rugby clubs seek regular protection for the winter months, not just for home games but also to cover the costs of travel and accommodation for wasted away journeys. Hunting, shooting and fishing enthusiasts are also regular customers of the insurance companies. Game shooting can fall foul of snow, high winds, pouring rain and reduced visibility. With a week's shooting on prime estates costing several thousand pounds, there is a strong demand for cover. Salmon fishermen take out policies against a deficiency of rainfall in the period before their expeditions. Elsewhere, organisers of regattas insure against boats becoming becalmed, hot-air balloonists against too little wind, and free-fall parachutists against excessive wind speed and low cloud.

It is not just promoters of events or participants who take out insurance. For example, when winter weather wreaks havoc on racing programmes the racecourse owners are not the only ones who stand to lose money if a meeting is called off. Caterers, car-park operators, local hoteliers and racing officials (such as starters and judges) all seek protection. A few years ago, British Rail (as it then was) bought a policy to cover against the cancellation of the Cheltenham Festival as it had laid on special trains.

Most insurers offer two types of policy against adverse weather conditions. The first provides cover if the event has to be abandoned altogether. As part of this line of business, some companies offer specialist cricket policies to insure against the possibility of no play or delays due to bad weather. Under the second type of policy payment is made if an agreed measurement of rain is reached during the period of time covered. If the amount of rain insured against is 5 mm and that is recorded, then the full amount of compensation is paid. However, to tempt the customer there is often an additional near-miss clause in which, in the instance above, 20 per cent of the sum might be paid if there is over 1.5 mm and 60 per cent if there is between 3 mm and 5 mm. With this type of policy organisers insure for the consecutive hours during which rain is most likely to affect attendance at their sports event. Often this includes the hours leading up to the start, as this is frequently a time when potential spectators decide whether or not to attend.

If the sum assured under such a 'Pluvius' policy – named after the Roman god of rain – is small, the organisers are often provided with a rain gauge and told to get the rainfall measurement themselves, though with the proviso that the person actually taking the measurement has no financial involvement in the event. The companies usually ask that an observer of assumed integrity, such as the local vicar or a policeman, monitor the reading. Careful instructions are given about placing the gauge on level ground not too near buildings or trees, with a warning to keep dogs and small boys away to avoid 'illegal' topping up! More generally, insurers rely on the 5,000 or so independent rainfall sites around Britain set up by the Meteorological Office and the Environment Office, which can provide an accurate record of how much rain has fallen in the area for the specified hours.

There are also specialist insurance policies dealing with winter sports. Skiers have insured their holidays and their bones for many years and such winter activities provide insurers with the bulk of their sports business. This is not surprising when it is claimed that about 5 per cent of the estimated 1.2 million British enthusiasts who go skiing or snowboarding each year face medical bills, and twice that number have something stolen whilst enjoying their snowy pursuits. Not all their injuries and losses result from the weather but, of course, skiing and boarding can only take place if the climate is right. Other clauses in winter sports insurance policies deal specifically with the elements, in particular whether there will be too much or, especially in Scotland, too little snow. Policies for no snow usually include clauses stipulating that bookings must have been made at least a month before the scheduled holiday.

The cost of insuring against the effects of adverse weather can be as variable as the weather itself. The key factors influencing premiums are the date of the event and its location. Thunderstorms with accompanying lightning are most likely in July and August whereas avalanches rarely happen outside the winter months. As for location, it is no use expecting a bargain rate if you are hosting a sports event anywhere near Seathwaite Farm in the Lake District, the Pen-y-Gwrd Hotel in Snowdonia or the Kinlochquioch Lodge in Lochabar near Fort William; for these three places share the dubious distinction of being the wettest inhabited sites in Britain, with annual rainfalls of around 3,120 mm. Geography, with its influence on the weather, helps determine the premiums. Generally they are higher in the western parts of Britain than in the east as there is more likelihood of rain there. However, there are differentials within this. In September 2000 Zurich Insurance, who along with Eagle Star dominate the weather insurance market, would have charged £283 for three hours of rain cover for a £30,000 event in Bristol, but £535 for the same cover in Keswick, simply because the Lake District is almost twice as likely to have rain at that time of year. Eagle Star, in the weather insurance business since 1920, charge sports clubs in Keswick – the wettest town on its rain map – almost 50 per cent more than Glasgow-based ones.

The small print must be read carefully. All insurance companies set a time limit before which the policy must be taken out. For summer policies there is generally a minimum two-week gap before the event, while up to six weeks' notice might be required in winter. This is partly designed to prevent clients insuring against a certainty such as a pitch already covered in deep snow becoming an unplayable bog when the snow melts. There are also limiting clauses similar to the excesses that all drivers face in their car insurance. These are often extended in order to reduce the premium. Spectators at Test matches and one-day internationals in England obtain a full refund if fewer than ten overs are bowled in a day. When the umpires abandoned the first day of the Lord's Test against Pakistan on 17 May 2001 without the toss taking place, it cost the England and Wales Cricket Board (ECB) around £400,000. Weather insurance had been taken out, but under the terms of the policy – which cost the cricket body £200,000 for the season – no compensation was to be paid until losses of £750,000 had been sustained. Winter sports insurance is particularly full of conditional clauses. Skiers who go off piste are not generally covered unless they stick to designated routes and are accompanied by a guide. They should not be tempted to do something different, such as snow-mobiling or ski mountaineering, and definitely not paraponting (parachuting off mountains) or heli-skiing (assisted by a helicopter), without agreement from their insurance company.

Some sports organisers opt not to take out insurance. The home nations rugby union authorities do not insure against cancellation of the Five (now Six) Nations Championship, presumably on the grounds that all matches will eventually get played and because, where England is concerned, there is an insatiable demand for tickets. The postponement of matches, rather than their total abandonment, during the foot-and-mouth epidemic justified their stance. Until the later 1980s the All England Club had not insured Wimbledon against bad weather. It then began to purchase ticket refund insurance. Under this policy tickets are fully refunded if there is no play and half the entrance fee is returned if it does not begin until after 6 p.m. However, the organisers did not have to claim for Wimbledon 2000 because, although matches were stopped by rain, there was sufficient play each day to prevent tickets having to be refunded. Matches in the Six Nations Championship and at Wimbledon can be rescheduled, one later in the year, the other further into the fortnight. On the other hand, cricket cannot easily rearrange a five-day Test match and insurance is now seen as vital. The first claim by the Test authorities was on 19 June 1987 when no play was possible on the second day of the match at Lord's against Pakistan. Further claims were also made for the fourth and fifth days! In its wisdom, however, the ECB later decided that the premiums were too expensive an investment, but they learned a hard lesson in 1991 when the Lord's Test against the West Indies was partly washed out. They had to pay out £400,000 in refunds, which meant that counties, which receive a cut from the profits of a Test series, lost about £20,000 each, a significant sum for county cricket. The ECB no longer attempt to do without weather insurance.

From the insurers' point of view, sports-related policies are just part of the general weather insurance market. As far as they are concerned, a rained-off tennis tournament is little different from a wedding turned white by snow or an ill-fated village fundraiser. Any weather condition can be insured against as long as there is weather data available on which to base actuarial calculations. Interestingly, the insurance companies have not raised premiums despite all foreboding of long-term climatic change. Their rates are based on an analysis of weather patterns stretching back over 100 years, so the odd bad month has little impact. On many of its policies, the rate per £1,000 of cover by Zurich Insurance has not changed since the 1950s.

Insurance companies, however, are not social services, and some clubs cannot get insurance. The New Road ground of Worcestershire County Cricket Club has flooded too often. Photographs of boats sailing across the square appear frequently in the local press – too frequently, according to the actuaries. As far as the insurers were concerned the incursion of 1998 was the last straw; it cost them £200,000 and since then the county has been on its own. The club now has to stand the risk itself. By 1990, restrictions had already been placed on the policy of the nearby Worcester racecourse company. When the track disappeared under a fathom of water that year, there was extensive damage to racecourse buildings and portable fences were washed away. Although most of the capital loss was covered, the labour costs had to be met by the racecourse executive.

Those who take out life assurance are usually happy when a claim does not have to be made. Similarly, Third Division Macclesfield Town willingly sacrificed the £12,000 that its directors paid to insure the third-round FA Cup match against Premiership side West Ham on 6 January 2002. The BBC had offered the club £265,000 for live coverage and, with average crowds of only 2,000, Macclesfield needed the money. In the previous year's third-round, non-League Dagenham & Redbridge lost a similar cash windfall when their replay against Charlton, scheduled for transmission on Sky, was postponed because of a frozen pitch. Macclesfield took no chances. Giant hot-air blankets were used to protect the pitch from the severe weather that forced ten other ties to be called off. Tons of grit and salt were spread on the open terraces so that spectator safety would not be an issue. And, just in case, the directors insured the match for £300,000. The premium was money that they did not mind losing because it brought peace of mind.

Sometimes there can even be a hidden silver lining to an insurance claim. In 1995, when all Boxing Day race meetings were called off because of freezing weather, Kempton actually made a bigger profit from the cancellation than it would have done had racing gone ahead. The racecourse had insured against a loss of its usual profit, but as there were doubts that racing would take place, some people would inevitably have been put off travelling to Kempton, and a smaller crowd than normal would have watched the meeting.

21

Human Elements

Spectators and groundsmen are not the only groups of sport buffs who keep a keen eye on the weather. The media, particularly the broadcast media, take a considerable interest in forecasts and on-the-spot conditions because of their impact on scheduling, transmission and commentary. Abandonments can result in rapid redeployment of outside broadcast units as cameras and personnel are switched to alternative sites. (Standby football matches or race meetings are held in reserve in case the main event is cancelled.) The technical quality and even the likelihood of transmissions can be affected by fog, wind, rain and storm. Who would be the guy at the golf tournament, apparently known affectionately as the 'mobile lightning conductor' – in other words, the technician who accompanies the roving reporter as he follows the action from hole to hole, toting a 'rucksack' from which protrudes a long, metal aerial? No chances can be taken with this equipment if there are storms in the vicinity!

Commentaries can also be affected by the weather. Essential notes have to be protected from unexpected hail showers or gusting winds; some seasoned professionals travel with a roll of Sellotape to tack down any wandering paperwork. Interviews conducted at wintry football stadiums emphasise the less glamorous aspects of media work. Many viewers will remember John Motson at Wycombe in 1990, facing the cameras in horizontal snow, huddled in cap and sheepskin coat. With crews braving the elements in exposed, unpleasant locations, and commentators often banished to cramped, precarious positions on gantries or the tops of grandstands – football commentator Alan Green regularly offers pithy remarks on the worst of these – the weather can have a major influence on the product enjoyed by the armchair sports fan.

If you listen to sport on the radio, you will appreciate many of these points. Football commentaries are often preceded by conversations between the presenter in the comfort of the studio and the team at the ground. Jokes about brass monkeys, forgotten gloves and hot Bovril abound. The present state of the weather and its impact on the playing surface are discussed. (As I listen tonight, our man at Lansdowne Road for the friendly between the Republic of Ireland and the United States refers to the pouring rain and the big puddles on the pitch. He thinks the game will be 'a bit of a slog'.) The wind strength is often referred to, can occasionally be heard as background noise and, in post-match summaries, is sometimes blamed for a poor spectacle. Fog swirling across stadiums has been known to reduce commentators to helpless laughter as they struggle to identify which player, or even which team, has the ball! They are the eyes of the listener and their job can be considerably hampered not only by misty or foggy conditions but also by rain sheeting down, snowflakes drifting across the ground or the glare of the sun.

At least modern radio broadcasters are not disadvantaged by the restrictions imposed on their forebears. The pioneering transmission of the Derby in 1926 was not allowed to offer any description of proceedings because of contractual arrangements with newspapers, and instead attempts were made to broadcast the atmosphere from Tattenham Corner. Unfortunately, the weather was terrible with softening ground and rain sweeping across the course. The normally boisterous crowd was intent on sheltering rather than shouting, the bookies had been silenced by the deluge and even the horses' hooves had been deadened by the wet turf. The broadcast was hardly a runaway success, unlike the winner, Coronach, who led from start to finish and won by five lengths, though in the second slowest time of the twentieth century.

Television has additional problems. Very strong winds can prevent the use of cameras high in the air which provide panoramic views of racecourse and golf course action from helicopters, airships or hoists. Camera crews sometimes have to face gale-force conditions, swinging in cages 100 metres above ground, to present coverage of events such as the Open golf championship. An outdoor camera brought in from the cold can suddenly steam up, ruining the picture, while lashing rain leaves its mark on lenses. This was not the only problem at Aintree in 2001 when torrential downpours brought difficulties for the BBC transmission of the Grand National as well as for horses, jockeys and spectators. Shortly before the start of the race, technicians found that the electrical supply to the Becher's Brook commentary box had been lost as a result of the atrocious conditions and it was too late to transport an emergency generator to the site. This left commentator John Hanmer with the unenviable task of calling the horses at Becher's from his position opposite the second fence: a difficult job with good visibility, a near impossible one at a distance of 800 metres in mud and rain.

There is also the problem, common to all live broadcasts, of filling those long hours at Wimbledon or Old Trafford when rain stops play. When Desmond Lynam presented the BBC coverage of the tennis championships, his ability to prattle on literally all day – broadcasts can stretch from 11 a.m. to 9 p.m. – was legendary. At least he had archive footage to fall back on. During the first televised match at Wimbledon in 1937, there was apparently a violinist on standby at Alexandra

Palace in case it rained! In his diary of football season 1995–96, BBC football commentator John Motson explained how bad weather can result in additional 'homework' if the scheduled game is postponed and another script has to be rapidly prepared. The hazards of driving in fog, snow and ice are a further aspect of a commentator's life that should not be underestimated. Like athletes, backroom staff, technicians and broadcast crews, they can travel thousands of kilometres a year in conditions which would keep sensible drivers at home, to deliver sport to the living rooms of the country.

The broadcast media can serve a useful function in disseminating information on postponed games, warning fans not to travel, and in summoning assistance to football grounds that require vast numbers of snow clearers; local radio can be particularly valuable in this context. On the other hand, the very presence of snow is one of the main reasons for the existence of Cairngorm Radio Ski FM, the local radio station for the Scottish ski resort near Aviemore, which provides up-to-date information about conditions. Radio 5 Live issues daily early morning bulletins about course inspections, cancellations and the state of the going for the day's race meetings, data which can also be accessed via Ceefax and similar outlets. The latter also offer brief daily weather forecasts for sports events and inshore waters, a service which BBC Radio 3 began in 1976.

Print journalism has none of the immediacy of radio and television and fewer weather problems in covering the sporting calendar. Where it excels is in crafting amusing or elegant descriptions of sport and the British climate. With considerable use of puns and wonderfully evocative phrases (one of our favourites, from December 1976, described the pitch at Leicester as 'like a prematurely iced Christmas cake') journalists create and enhance for readers the weather pictures available on television. Weekend newspapers, like regional radio and television, have also expanded their coverage of leisure-related weather forecasts in recent years, offering specific reports for skiers, sailors and hillwalkers. These are sports in which lengthy periods outdoors bring with them the issue of safety in what is often a rapidly changing British weather map. Avalanche risks in the Scottish mountains are also assessed.

Everyone connected with sport – players, administrators, promoters, spectators, groundstaff and broadcasters – pays attention to the weather forecaster, the linking human element in the relationship between this form of outdoor entertainment and the climate. With ever more sophisticated radar, satellite and computer predictions, there are increased expectations (fond hope or foolish daydream!) of accurate weather forecasts; to tell supporters what to wear for maximum comfort, to warn groundstaff to get the covers ready, to inform event organisers and broadcasters of possible problems so that contingency plans can be made. Some sports venues, such as Wimbledon and Old Trafford, now have their own weather centres, others rely on hotlines to the local meteorological office. The All England Club has two radar systems which supply short-term forecasts and can predict when and how long it will rain, but the head groundsman still contacts the Met

Office three or four times a day for more detail. The Professional Golfers Association tour in the United States even has its own full-time meteorologist on site to monitor potentially dangerous weather events such as tornadoes and lightning strikes or to give caddies predictions on wind speed and direction.

Shipping forecasts are a longstanding feature of BBC radio, and forecasts for coastal waters are also broadcast regularly. Their importance for weekend sailors cannot be emphasised enough. More recently, regional television has included predictions for sailors and walkers in weekend forecasts and even general weather bulletins often contain a brief reference to expected conditions at major sports events, both at home and abroad. If your team are playing in Europe or you fancy a trip to your local racecourse, information about the likely weather has now become part of the regular weather slot, a reflection, perhaps, of the central importance of sport in the national culture.

Several outdoor activities can now benefit from an increase in specialist weather services. For skiers in Scotland, there are separate hotline numbers for each resort, giving updates on snow conditions. There are also commercial snow information services, road access reports and both general and individual summaries of weather prospects, together with a Ski-Scotland web site. Local telephone and fax reports suitable for rambling clubs and walkers are now available, as well as more detailed forecasts for climbers in the Scottish Highlands, Snowdonia and the Lake District, which include items on cloud base and freezing levels. Small radios and mobile phones are compact enough to be carried on expeditions and used to pick up warnings of imminent changes in the weather, hopefully before they arrive!

Early warning is also vitally important for groundstaff who need to know whether to cover football pitches in winter (to protect against snow, frost and ice) and cricket pitches in summer (to shield them from rain). Racecourses cannot cover lengthy tracks but still require accurate assessments in the days before a meeting if they are to provide the best possible surface. Watering may be sanctioned if the ground is firm and no rain is forecast. If unexpected showers occur after watering, good going can rapidly turn soft and the Met Office will not be popular with the clerk of the course, who in turn will incur the wrath of trainers and bookies. And, of course, forecasters sometimes get it wrong.

When the Irish Rugby Football Union (IRFU) reluctantly called off the Five Nations fixture against England in a snowy mid-January in 1985, pitch covers were in place and the visiting team had already arrived in Dublin. On the Thursday evening before the game, the authorities had been optimistic that it would go ahead on Saturday as planned. The IRFU, however, was persuaded by forecasters that the weather would remain severe and, with safety and travel for spectators a key issue, the match was postponed on Friday morning. Then the snow turned to sleet . . . and the sleet turned to rain. Dublin airport re-opened, the lying snow began to thaw and there was criticism in the press that the decision had been taken too early. Frustration, financial cost and time-consuming rearrangements were the consequences of a forecasting *faux pas*.

Fans are often inconvenienced by forecasting errors but most complaints come after late postponements. When supporters of the Glamorgan cricket team travelled to Lancashire for an AXA Life Sunday League game in 1997, they found that the match had been called off only an hour and a half before the scheduled start, although it was a sunny day with a good breeze. The groundsman had decided earlier in the week that the whole square needed watering and, with rain forecast, had left it open to the elements. After the rain had done its job, he covered the pitch, but unfortunately the unsettled spell continued far longer than expected and the ground didn't dry out. To make matters worse, the Australians were scheduled to start the third Test against England on the following Thursday. The Sunday match became a casualty of the weather as no one was prepared to risk the damp pitch turning to mud before the international.

Getting it right can be particularly tricky in the case of motor racing circuits, which are potentially dangerous if the forecast is wrong. The size of the track and the existence of microclimates can lead to situations in which one side of the circuit is bone dry and the other awash with rain. Spa in Belgium is notorious for this phenomenon and its rapidly changing weather patterns, while Estoril in Portugal is renowned for its unpredictable Atlantic winds. If the weather is unsettled and the forecast uncertain, drivers face difficulties over tyre choice. If the track is wet but the weather forecasters predict dry conditions, decisions have to be made about whether to start on wet tyres – safer but slower – or untreaded tyres – fast and suitable for good weather but riskier on still damp but drying surfaces. The start of the 1991 Spanish Formula One Grand Prix in 1991, held at the Circuit de Catalunya, Barcelona, presented exactly this dilemma. The track was wet but rain had stopped and the warm-up took place under grey skies. In spite of a good forecast, all the drivers opted for wet tyres. (Alain Prost allegedly chose slick tyres but was overruled by his team management!) Ten minutes after the start it began to rain again, a case of Formula One team 1–Weather forecasters 0.

However it turns out, the poor forecaster gets blamed for the weather. In January 1988 an unexpectedly heavy snowfall led to road deaths, cases of hypothermia, stranded sheep and cancelled sports fixtures. There were complaints from police and public alike that there had been insufficient warning. The Met Office was quoted as saying, 'We have been forecasting snow for the past few days. The volume took us by surprise but it's just normal snow.' However, forecasts are improving. The launch of new satellites in the first five years of the twenty-first century will produce better data, which is particularly good news for fans of winter games. Computer predictions, according to the president of the Royal Meteorological Society, are at their most accurate in winter – bad luck for supporters of cricket!

ALWAYS TAKE THE WEATHER WITH YOU

22

Brits Abroad

ritish athletes have competed overseas for more than a century. English jockeys frequently rode in Europe in the 1860s, England cricket teams toured Australia in the 1870s and British rugby teams began to visit the southern hemisphere in 1888. There were British representatives at the first modern Olympic Games in Athens in 1896, Britain and America held the inaugural Davis Cup tennis match in Boston in 1900 and the England football team played its first match on foreign soil nearly 100 years ago, in Austria in 1908. British sportsmen and women sometimes encountered weather patterns similar to home, but on other occasions they found themselves facing extremes unheard of in north-west Europe.

Brits tend to have few problems when presented with pouring rain and gale-force winds – they regularly deal with such atrocious conditions in their own backyard. Peter Buckley, from the Isle of Man, won the gold medal in the 120-mile (190-kilometre) cycle road race at the 1966 British Commonwealth Games in Kingston, Jamaica, in spite of heavy rain and wind. As the wild weather swept leaves, twigs and stones across the course, causing 11 riders to skid out of control, he managed to set a record time for the championship of 5 hours 7 minutes 52.5 seconds. The rain, he said, was 'worse than Oldham'. When the golf World Cup took place over four days in Hawaii in November 1987, the tournament was plagued by storms. Funnily enough, the teams that coped best were Scotland and Wales, who played-off on a day of torrential rain and high wind. Ian Woosnam, for the victorious Welsh team, said that he had never played better in worse circumstances. And when the final of the President's Prize shooting competition was staged at Bloemfontein, South Africa, in April 2001, driving rain forced the organisers to abandon the two-range final in favour of a single shoot over 900 metres. In trying conditions it was Wing Commander David Calvert of Northern Ireland who held his nerve, scoring a bull's-eye in all but one of his 15 shots, to win the event.

But playing away, particularly in the more far-flung corners of the world, often involves adjustments to alien food and drink, to unfamiliar pitches and to climates very different from that of the British Isles. Living, training and competing for much of the year in cool weather, British athletes can find the process of acclimatisation to searing heat a painful and, sometimes, a dangerous one. Although they have usually responded heroically to soaring temperatures and tropical humidity, their efforts have met with mixed results. When Colin Montgomerie played in the US Open at Oakmont Country Club, Pennsylvania, in June 1994, he had to endure four days in temperatures of over 32°C, with high humidity. Wearing a straw hat for protection, Monty fought his way into a play-off for the title with Loren Roberts of the United States and Ernie Els of South Africa, but the weather was debilitating, particularly for European players. Bernard Langer admitted to being close to collapse at the end of an earlier round and Monty's caddy needed treatment because of the heat. The three-way play-off, held over a further 18 holes on an equally hot day, resulted in poor golf by all the contestants. Under pressure from the climate or the occasion, Montgomerie carded 78, the other two 74. Els finally won on the second hole of sudden death.

Home nations' rugby teams often experience difficulty in adjusting to conditions in the southern hemisphere. When the Great Britain rugby league team played Australia in June 1977, the temperature in Brisbane was 26°C. It was noticeable that the British forwards slowed up during the second half, almost certainly because of the heat, and Australia won 19–5. Switching codes, September 1969 saw Scotland lose 20–6 to Argentina on a hard pitch in hot weather, the Scots wilting in the first half and visibly tiring in the second. Two weeks later, in lower temperatures and on a softer pitch, they won 6–3. On a tour of the Far East in 1977, they managed to overcome high humidity and a weak opposition in Thailand by strategically placing water and fruit on the touch-lines. But in June 1993, against a strong Western Samoa side in sweltering heat, the Scots were beaten 28–11. It isn't easy to play 'the great and noble winter game' in average temperatures twice as high as Jedburgh in January or Melrose in March!

British footballers have encountered similar problems. Manchester United, playing in the Club World Championships in the Brazilian summer of 2000, were warned by local hero Pele not to underestimate the impact of the weather. Rio in January bakes in 40-degree heat and oppressive humidity, and the players must have realised what they were up against on their first visit to the dressing-rooms of the Maracanã Stadium; individual benches were equipped with oxygen facilities for use at half-time and after the games. The manager certainly knew the problems. Alex Ferguson had taken the Scotland team to the Mexico World Cup in 1986 and seen players lose as much as 5 kg per game in a similarly hot and humid climate. United, controversially opting out of the British winter and their defence of the FA Cup, found themselves instead with a schedule of 22 hours flying and possibly 4 fixtures in 10 days of scorching summer. In some ways, it turned out to be less onerous than they feared. In spite of the latest medical advice, early-morning training sessions to avoid the heat and match tactics which involved slow build-up and play, they were eliminated after a draw and a defeat, and were soon on the plane home to murky Manchester.

Jack Charlton, another veteran of World Cup campaigns as both player and manager, has consistently warned of the dangers to northern European teams of playing in the cauldrons of overseas stadiums. His Irish team, all of whom played their football in British leagues, were sorely disadvantaged during the 1994 finals in America by temperatures of over 32°C and inadequate pitch-side water supplies. After their first game against Italy in New York, which they won 1–0, Tommy Coyne of Motherwell is said to have collapsed with heat exhaustion on the return trip to the Irish team base in Orlando, Florida, where they had been trying to acclimatise for two weeks. Henning Berg of Blackburn Rovers, playing for Norway, one of Ireland's group opponents, admitted to drinking 7 litres of water and still losing 4 kg in the 35°C heat of Washington DC in a match against Mexico. Charlton made several formal complaints to FIFA, accusing them of indifference to the plight of players who normally performed in European winters. They rewarded him with a reprimand. The American public was more generous, expressing admiration for the fitness of footballers who played for 90 minutes in extreme conditions without time-outs, particularly as at least 40 spectators per game were being treated for the heat. The *Sunday Times*, supporting Charlton's view, wondered who would be blamed if a player collapsed in the ferocious temperatures, and castigated FIFA for their lack of concern about player welfare. Unlike his counterparts in sports administration, Charlton, together with all late twentieth-century coaches and competitors, was well aware of the health and safety issues involved in playing sport in hot weather. Unfortunately, this was not always the case.

Weatherbeaten!

On 7 August 1954, the sun beat down on a packed stadium for the last day of the fifth British Empire and Commonwealth Games in Vancouver, Canada. Some 35,000 spectators had turned up to see the final, climactic events and within the space of 15 minutes they witnessed two great dramas, one triumphant, one tragic. Firstly, they saw Britain's Roger Bannister battling past the Australian, John Landy, to win the 380-yard event, both men breaking the four-minute barrier. Moments later, they saw Jim Peters, captain of the England team, enter the stadium as the frontrunner in the marathon, with only 352 metres between him and the winning line.

He never crossed it. Weaving across the track like a drunken man, legs buckling, arms hanging limp, he fell to the ground only to stagger to his feet and resume his tottering gait. After a few metres, he collapsed again . . . and again . . . and again. He was the victim of dehydration and severe heat exhaustion, brought on by too fast a pace on the hottest day of the Games. To the horror of the watching crowd, the gallant athlete repeated this macabre performance 11 times in all before officials finally helped him from the track. He was taken to hospital and made a rapid recovery but never ran a marathon again. The heat – some said it reached 32°C – took its toll of the entire field, with only 6 of the 16 competitors completing the distance. The winner, Joe McGhee of Scotland, also finished in a distressed state after falling several times on the road circuit. Unlike Peters, however, he had been mentally alert enough to rest along the way, and he eventually crossed the winning line in a time of 2 hours 39 minutes 36 seconds, over 20 minutes slower than the world record set by the Englishman back home.

Thirteen years later, a more calamitous episode captured the headlines of the sporting press. Tommy Simpson was Britain's top professional cyclist. He rode himself to death in the Tour de France in July 1967 in the unrelenting heat of Provence, a victim of weather and sporting ambition. As an amateur, he had won bronze in the Melbourne Olympics and silver at the Cardiff Empire Games. As a professional, he won the Tour of Flanders in 1961, became the first Briton to wear the coveted yellow jersey in the Tour de France of 1962 and was world road race champion in 1965. Sidelined through injury for most of 1966, the 29-year-old saw 1967 as perhaps his last chance to win the blue riband event.

During the 13th stage of the Tour, having already covered 140 kilometres from Marseilles in temperatures of over 35°C, the cyclists had to ascend Mont Ventoux, 1,890 metres high and treeless. It may have reached nearly 50°C on the afternoon of 13 July 1967 on the barren, exposed mountainside. Simpson, among the leaders at this stage, was seen to drop back as the riders toiled their way upwards. He collapsed with breathing difficulties several kilometres from the summit, rapidly lost consciousness, and was flown by helicopter to hospital in Avignon, where he died. The cause of death was given as dehydration and exhaustion, but the autopsy also revealed traces of amphetamines and alcohol, which probably contributed to the eventual tragedy. The British cyclist was not alone in resorting to stimulants. Even 35 years ago, there were suspicions that the physical demands of the Tour spawned a drug culture amongst participants in this most gruelling of endurance sports. Tommy Simpson had simply wanted to win, but he pushed himself beyond his limits on a scorching hot day – and paid the ultimate price.

Barry McGuigan of Northern Ireland thought he'd got it right. The WBA featherweight champion was scheduled to fight Texan Steve Cruz in the heat of Las Vegas in June 1986 and he was 5–1 on favourite to retain his title. His preparations included a month at Palm Springs in the California desert to acclimatise and train for 15 rounds in temperatures of over 38°C. When the contest began, in an outdoor arena specially constructed in the parking lot of Caesar's Palace casino, McGuigan had the support of over 2,000 fans chanting his name, while his opponent, ranked ninth in the world and a plumber by day, could muster fewer than 100. But the fight didn't go according to plan. It was 45°C when the boxers stepped into the ring and, although in the early rounds Cruz was described as little more than a sophisticated punching bag, he fought back, knocking McGuigan down in the tenth. Going into the 15th round, the Irishman was still ahead on points, but his reserves of stamina finally gave out, allowing Cruz to floor him twice. He got to his feet on both occasions, but it was all over and he lost the fight on a unanimous points decision. He left the arena on a stretcher and was later taken to hospital, attached to a drip, and kept in overnight. Even his opponent, accustomed to fighting in the fierce heat of western America, admitted that the sweltering sun had troubled him in the early rounds.

With hindsight, McGuigan said that four weeks heat training had not been long enough, and that he had not been as sharp as in his previous title fights, all held in Britain and Ireland. By the seventh round, the balls of his feet and his big toes had blistered on the scorching hot canvas of the ring and, in retrospect, it had been a mistake to arrange the fight for just after 6 p.m. local time to accommodate British television audiences. It had been too early and too hot. The chief medical officer of the British Boxing Board of Control agreed, questioning the wisdom of allowing McGuigan to box outdoors in the high temperatures of early evening before the sun had set, and expressing concern at the levels of exhaustion and dehydration experienced by the boxer. It was nearly two years before Barry McGuigan fought again. In the heat of Las Vegas, he was beaten as much by the weather as by his opponent.

In all these cases, the timing of the event was a factor in the final outcome and lessons have been learned. Marathons at major championships are often scheduled to start in the cooler morning, sometimes as early as 6 a.m. to avoid the worst of the heat. After Simpson's death, the Mont Ventoux stage of the Tour de France was held later in the day so that riders were not faced with the exhausting climb in early afternoon. Rescheduling alone, however, cannot remove the element of risk in the most physically taxing sports. Only three years after Simpson's collapse, the two leaders Eddy Merckx and Martin Van Den Bossche were rushed to an ambulance to receive oxygen on completing the ascent.

Hot Horses

It isn't only humans who suffer from heat exhaustion and dehydration. When the three-day event World Championships were held in Lexington, Kentucky, in 1978, it was the horses that found it difficult to cope with the sultry conditions. Warrior, ridden by Jane Holderness-Roddam, had won the Badminton Horse Trials that spring and was picked for the British team to America. But two-thirds of the way round the cross-country course, he began to stagger, and he and his rider collapsed into a water hazard. Although they carried on for a further eight fences, they were eventually eliminated and, according to his rider, the horse never really recovered from the experience. In spite of a third place at the Burghley Horse Trials in 1980, he was retired from major competitions.

Much controversy surrounded the Lexington event, with tales of exhausted and disabled horses on the one hand, and riders and officials insensitive to the climate on the other. The debate about whether horses should be asked to gallop in heat and humidity has subsequently been fuelled by the choice of Olympic venues – Los Angeles, Seoul, Barcelona and Atlanta have been far from ideal for three-day eventing. Some riders, notably those from New Zealand, consistent medal winners since 1984, have openly voiced doubts about the hot, hard and uncomfortable conditions at the Olympics, pointing out that their sport is best suited to temperate climates. Even in northern Europe the unpredictable weather can cause problems, however. When the World Equestrian Games were held in The Hague in 1994, they coincided with the hottest July in the Netherlands since 1756. Several horses finished the speed and endurance phase clearly exhausted, including King William, ridden by Mary King for the Great Britain team. Another Briton, Caroline Sizer, competing as an individual on Ghost Town, was not allowed to start the cross-country when the horse's temperature failed to come down after the second phase of roads and tracks. The horses recovered quickly, but the incidents focused attention on animal welfare at the forthcoming Atlanta Olympics in 1996, where conditions were expected to be gruelling.

In the end, the horses competing at Atlanta finished no more tired than at a British event, partly because temperatures were lower than expected on the day of the speed and endurance phase, and partly because of modifications to the tests to combat the climate. The animals had also been given three weeks to acclimatise to conditions, and the latest techniques had been used to cool them down at the enforced breaks on cross-country day, including fans blasting atomised water. None of this was any help to the British team, who finished fifth.

Hard Luck

It isn't always hot and sunny abroad. Occasionally, Brits are beaten by the same sort of weather they might find at home, or perhaps a more extreme version. The arrival of rain and wind can herald problems for athletes the world over and cause winning situations to slip away, as the England cricket team discovered in 1990. Their tour of the West Indies had started very promisingly, with Graham Gooch leading them to a first Test win against the home side for 16 years. Then the weather intervened, and the second Test at Georgetown, Guyana, was a complete wash-out, only the fifth in the history of Test cricket. In the third Test at Port of Spain, Trinidad, England were in a commanding position by close of play on the fourth day, requiring only 151 in their final innings with a whole day ahead. They had reached 73 for 1 halfway through the morning session and victory seemed likely when chance dealt two swift blows: firstly, Graham Gooch broke his finger . . . and then it rained.

When the teams finally took the field again after a three-hour break, it was a race against the light. England only required 78 from 30 overs, but a combination of cynical time wasting by the home side and darkening skies overhead put paid to their chances of becoming the first England team since 1957 to win successive matches against the West Indies. They soldiered on to reach 120 for 5 but with West Indies seam bowlers, Ambrose and Bishop, thundering down in rapidly fading light, conditions were becoming too dangerous to play on. David Capel rejected the first offer of light on instructions from the England balcony, but 15 minutes later he and Jack Russell were forced to retire, needing only 31 runs to win. 'Never,' said past England captain, David Gower, 'will a West Indies side have been more grateful for the intervention of the rain gods. The weather had turned the near certainty of victory, and at worst a squared series, into a draw, and what followed was predictable enough. West Indies won the fourth Test by 164 runs, the last by an innings and 32 runs, and the series 2–1. No wonder England felt aggrieved and blamed the rain for coming between them and a successful overseas tour.

The fates also conspired in many ways against the British Lions tour of 1977, and one of the crucial elements in their defeat was the weather. Hampered by great expectations, they were favourites to beat the All Blacks after the touring teams of 1971 to New Zealand and 1974 to South Africa had returned victorious, the first conquerors of southern hemisphere opposition. The Lions were weakened before the tour even began by the loss of top players Gareth Edwards, Gerald Davies and J.P.R. Williams, who were unavailable, and during it by injuries to several key personnel. There were problems with coaching and management, with an acrimonious New Zealand press and hostile public and even with domestic issues such as accommodation. Above all, hanging over the whole three-month venture was an appalling winter.

It rained continuously. Even if this is not strictly a meteorological fact, it was the overriding impression left in the memories of the squad and the minds of their fans back home. The 'All Black' pun was used in so many photo captions that drenched, mud-covered players, so coated in slime that their shirts were indistinguishable, came to epitomise rugby that year. (Like the 1992 film *The Piano*, set against a backdrop of seemingly endless torrential downpours and trench-like ooze, it can have done nothing to advance tourism in New Zealand.) The persistent rain led to sodden, heavy pitches, which in turn made it difficult to play good rugby. Training in the rain day after day, trying unsuccessfully to dry out wet kit, whiling away long hours indoors because the weather precluded any swimming, golf or other outdoor leisure, all had a cumulatively demoralising effect on squad members as the weeks dragged by.

In spite of these difficulties, the Lions won 21 of 25 matches but, unfortunately, three of the defeats were in the Test series. The first Test, at Wellington, took place in strong winds on a pitch that had been partially dried out by helicopter after a night of continuous rain. The All Blacks won 16–12. The Lions grabbed a result in the second Test, winning 13–9, and lost the third 19–7. They desperately wanted to win the final Test to square the series, and looked like doing so until the dying minutes of the match. Then, leading 9–6, they were caught out by an All Black try in injury time, and lost 10–9. The Lions' manager, George Burrell, reckoned that the tour had been marred by 'the filthiest, wettest winter in years', with match after match spoiled by pouring rain and waterlogged grounds. Luck had not been on their side.

Paula Radcliffe won her event but lost a record-breaking opportunity because the weather refused to cooperate when she was competing in the European 10,000 Metres Challenge in April 2001 at Barakaldo in the Basque region of northern Spain. Her recent times had been so good that she had high hopes of beating the 15-year-old European record set by Ingrid Kristiansen in Oslo in 1986. She won convincingly, lapping the entire field with the exception of Irina Mikitendo, an Olympic 5,000-metre finalist, but her record chance was swept away by gale-force winds off the Bay of Biscay. Not only did it blow her around on the track, it whipped sand up from the long-jump pit into her eyes on every lap. With such difficulties, it was not surprising that her winning time was 42 seconds slower than Kristiansen's. She said afterwards that the 25 laps were the most gruelling she had ever contested.

Against All Odds

There have been instances in sport when the weather has been hard to contend with, almost impossible to overcome, but on that particular day, one competitor has mastered it better than the rest. It happened to Tony Jacklin on the first day of the US Open in June 1970. Jacklin, the 25-year-old holder of the British Open title, took a two-shot lead for a round of 71 while the world's best golfers struggled to combat the conditions. The course at Hazeltine, Minnesota, was the second longest in US Open history and had been a controversial choice. Its reputation was not enhanced on the opening day of the tournament when a 65 kph wind wrecked the chances of Arnold Palmer, who shot a 79, Gary Player (80) and Jack Nicklaus (81). Balls were being caught by the gale and whirled out of bounds, and 20 of the first 30 starters recorded scores of 40 and over for the first nine holes. Jacklin made the turn in 33. The American golfer, Mason Rudolph, said he had seldom experienced such difficult conditions, and those had been in Britain, which perhaps helped to explain why the only Englishman in the field topped the leader board. Jacklin remained unperturbed. Increasing his lead in the second and third rounds when the weather turned calm and sunny, he shot a 70 on the final day to win the event by seven strokes from American Dave Hill. He was the first Briton to win the US Open for 50 years, emulating the success of Ted Ray in 1920.

When British Formula One racing driver James Hunt lined up on the grid in October 1976 for the first Japanese Grand Prix, the final race of the year, he was two points behind world champion Niki Lauda. It had been a dramatic season, dogged by contentious disqualifications to Hunt and overshadowed by Lauda's near fatal crash at the German Nurburgring. Lauda had been leading the driver's championship at that time and, only 12 weeks after his accident, he was on the grid at Fuji in a last gasp effort to defend his title. The weather was atrocious, with torrential rain and mist forcing a 90-minute delay. With visibility down to 100 metres and puddles on the track, many drivers were unhappy about starting the race. But start it did, largely because the world championship was at stake and over 70,000 spectators had turned up to watch. Hunt, racing from the front row, was able to avoid the blinding spray thrown up by the following cars, but Lauda was not so lucky and, with his vision obscured by huge plumes of water, he retired within the first two laps, describing conditions as 'suicidal'.

Hunt, meanwhile, had to finish no lower than fourth to win the championship, and he led around the rain-soaked circuit for much of the race. The drama, however, was not yet over. The deluge stopped after 30 laps had been completed, but the track was still under water in places and highly dangerous. Hunt, slithering around flooded corners on worn tyres, eased up and dropped to third in the hope of nursing the car home without a pit stop but, on the 69th lap, one tyre gave out completely, and he was forced into the pits. When he rejoined the race, he was in fifth position with the championship seemingly lost, but a combination of good driving and good fortune enabled him to claw his way back. He eventually took the chequered flag in third place, and the Formula One title by a single point.

In Gothenburg, Sweden, on the evening of 11 May 1983, it was the underdogs who triumphed against all odds. Aberdeen FC, managed by a youthful Alex Ferguson, were playing for the first time in a European final, the Cup-Winners' Cup, against Real Madrid, veterans of ten previous European finals. Although untried in Europe, the Aberdeen team sheet now reads like a list of top Scottish stars and proven managers – Jim Leighton, Willie Miller, Alex McLeish, Mark McGhee and Gordon Strachan. It was Eric Black, however, recently appointed assistant manager at Coventry, who opened the scoring in the sixth minute from a Strachan corner.

According to *The Scotsman*, it was 'an unbelievably wet night'. There had been torrential rain all day, together with thunder and lightning, and at one stage the fixture looked to be in doubt. When the match kicked off, virtually the only spectators were the 10,000 drenched but undeterred Aberdeen fans. The playing surface was treacherous, but after Real equalised with a penalty in the 14th minute, they looked the happier of the two teams. Although Aberdeen might have been expected to cope better with the appalling weather, they were finding it difficult to pass accurately on the saturated pitch.

As the second half and then extra time wore on, the rain continued to sweep across the Ullevi Stadium, and the conditions made enormous demands on the stamina and character of both sets of players. When, however, substitute John Hewitt, the Scotland Under-21 international, scored the winner midway through the second period of extra time, it was no more than the Dons deserved. They had been hampered all night, not only by the elements, but by a referee who had repeatedly turned down appeals against hefty challenges and had refused at least one good penalty claim. But the weather had not finished with the team yet. Their flight home and triumphant procession through the streets of Aberdeen was almost delayed: by fog!

Twelve Tall Tales (or What Some Brits Discovered Abroad

The fact that a third of the following relate to cricket must say something. It probably demonstrates that any game requiring five days of uninterrupted dry weather is asking for trouble!

1. Cold Cricket: The journal of Capt. (later Admiral Sir) William Parry's second voyage through the frozen wastes of northern Canada in search of the Northwest Passage was published in 1824. The expedition was a failure, with the way forward for his two boats blocked by ice, but time among the ice floes was not entirely wasted. The book shows a drawing of a cricket match played between the ships' crews on an icy 'pitch' within the Arctic Circle. The temperatures were said to be more than 30 degrees below zero.

2. Burning in the Wind: In 1927, Sir Henry Segrave, one of the pioneer British racing drivers, retired from the sport to concentrate on breaking the world land speed record. In a 1,000 hp red Sunbeam, he made his attempt at Daytona Beach, Florida, in March of that year, and clocked over 331 kph on his first effort over the course. The wind pressure made steering extremely difficult and, after his record run, he complained of windburns and aching wrists, caused by trying to hold the four-ton car steady on the sands. His official world record, averaged over two runs, was given as 326.06 kph.

3. It was not Football Weather: When England lost 4–3 to Spain on 15 May 1929, it was their first defeat abroad. They had asked for the match in Madrid to be played later in the day, but the Spanish football authorities insisted on a mid-afternoon kick-off. England started brightly enough and led 2–0 at half-time, but they were unaccustomed to the sweltering conditions and, after a hard domestic season, the players began to wilt in the second half. One newspaper reported that they were drenched in perspiration, with beads of sweat dripping from their chins, and described the heat as 'torrid'. The Spaniards, coached by former Middlesbrough and England winger Fred Pentland, capitalised on their opponents' weariness and emerged the winners, heralding their arrival on the international soccer scene.

4. The Last 'Timeless' Test: The rules were different then. England's last tour of South Africa before the war had resulted in three drawn games and one English victory and, with only one match separating the sides, the final Test was supposed to be fought to a finish. It started at Durban on 3 March 1939. By the end of the sixth day, South Africa had posted totals of 530 and 481, while England were all out for only 316 in their first innings. This left them needing to amass 696 for victory, nearly 300 more than any side had previously made in the last innings to win a Test match. On the seventh day, they scored 253 for the loss of only one wicket. On the eighth day, it rained and no play was possible.

On the 9th day, with the pitch improving, they reached 496–3 and by the tea interval on the 10th, they were only 42 runs short of a famous victory. Then the rains came, not a drizzle, but a downpour and no further play was possible. By the morning of the 11th day, the England team had already embarked on a two-day rail journey to the coast to catch a homeward-bound ship, leaving a drawn match and a number of first-class records in their wake. With victory so close and nine days spent in the field, it seems incredible that time could not have been found to play to a finish, or an alternative passage secured for the visitors. It signalled the end of the 'timeless' Test, the longest match in cricketing history and one that would never be repeated. (This was not the only occasion that a 'timeless' Test had to be abandoned to let an England team catch the boat home. In April 1929, the final match of four against the West Indies, at Kingston, Jamaica, was halted by rain on the eighth and ninth days and declared a draw, leaving the series all square.)

5. The Barefoot Marathon: Jack Holden was 43 when he won the marathon at the Amateur Athletic Association championships in July 1950 in a personal best time of 2 hours 31 minutes 3.4 seconds, his fourth marathon victory of the year. His first, at the British Empire Games at Auckland, New Zealand, was memorable for the deluge in which it was run and the dog that nearly disrupted the race. Holden was leading the field in pouring rain when his running shoes became waterlogged and split. Although water was lying 150 mm deep in some parts of the road, he decided that it would be easier to run barefoot for the last 13 kilometres of the race. As if this was not hard enough, he was attacked by a dog in the final stages – either a Great Dane or a poodle, according to different versions of the story. He finally entered the stadium, his feet covered in cuts and bruises, but he beat the rest of the field, including the Olympic silver medallist, by over four minutes.

6. Sun Stops Play: The third Test against New Zealand at Christchurch in March 1963 saw Fred Trueman overtake Brian Statham's world record of 242 Test wickets and it also featured a rare incidence of 'sun stopped play'. With England batsmen David Sheppard and Ken Barrington at the crease, the early evening light was reflected off an aluminium grandstand roof, causing the batsman, wicketkeeper and fielders at the striker's end to be blinded by the glare. Barrington appealed against the conditions and play was halted ten minutes early.

7. Rain Stops Play: On the first day of the third Test between England and Australia in 1971, England won the toss and put the home side in to bat. But before the teams emerged, the heavens opened over Melbourne and three days of continuous rain followed. The match was abandoned and it was agreed that a one-day game should take place on what would have been the fifth day of the

Test. England were all out for 190 in 39.4 eight-ball overs, and Australia won by five wickets. The one-day international was born.

8. The Water Polo Test: On 14 June 1975 no game of rugby should ever have been played. Sections of the Eden Park ground in Auckland, New Zealand, were under water and the fire brigade had been called in to pump the worst of the deluge from the pitch after 100 mm of rain had fallen in 12 hours. It was still raining when Scotland and the All Blacks agreed to play the match, largely because the visitors were flying home next day and 45,000 spectators had turned out in appalling conditions to see them. Scotland battled hard against driving wind and rain in the first half and were only 6–0 down at half-time. However, in a cruel meteorological quirk of fate, the wind swung 180 degrees during the interval, and the New Zealanders ran out 24–0 winners. The team captains acknowledged the farcical match conditions, Leslie of New Zealand referring to 'one of the greatest moments in New Zealand swimming' and McLaughlin of Scotland remarking that it was 'sheer luck that nobody drowned'.

9. Not Weatherbeaten: Max Telford, a native of Hawick in Scotland who had emigrated to New Zealand, became the first man to run through Death Valley, California, one of the hottest places on earth. Between 2 and 4 of July 1976, he ran the 384 kilometres there and back in a temperature of 49°C in the shade. The ground was so hot that a member of the camera crew filming the event was apparently able to fry an egg. Telford normally ran 320 kilometres a week in training, but his acclimatisation programme to the heat of California also included regular exercise on a treadmill in a laboratory heated to 54°C.

10. Fewer Fags: At the start of the 500 cc world championship season in 1977, Barry Sheene, the title winner in 1976, embarked on a rather less orthodox fitness campaign in order to conquer the heat and humidity of Venezuela. Remembering that the previous year had seen temperatures of 3°C at the San Carlos circuit, he started a daily two-hour exercise programme, gave up alcohol, cut down smoking and arrived in the country five days early to acclimatise. He won the race, later retaining his world title.

11. Water at the Creek: The US PGA championship takes place in August, often in mid-western or southern states such as Ohio, Oklahoma, Indiana or Kentucky, where the heat and humidity can be overwhelming. The unhelpful timing and uncomfortable climate, with temperatures usually soaring above 35°C, are sometimes cited as factors in Europe's failure to capture the event, but this lack of success was not the result of poor preparation in the case of Nick Faldo. In 1990, having already won the US Masters and the Open at St Andrews, he was bidding to become the first golfer since Ben Hogan in 1953 to win three majors in one season. Before the tournament, held at Shoal Creek, Alabama, Faldo went

to considerable lengths to combat the anticipated dehydration of playing in the Deep South, although his approach was different to that of Barry Sheene. In the fortnight between the Open and the US PGA, he deliberately increased his intake of water, putting on nearly 3 kg in the process. It didn't appear to help him. Opening with a 71, he carded 75 in blistering heat on the second day and 80 on the third. He finished the tournament tied for 19th place, his worst finish in a major championship for over two years. After admitting to extreme tiredness, he was advised by a specialist to rest for the next two weeks.

12. Oxygen at the Ready: When Great Britain played Papua New Guinea at Lae in September 1996 in the first rugby league Test of a series, it was just as well that the unlimited interchange rule was in force. It enabled coach Phil Larder to make a vast number of substitutions to combat the stifling heat and humidity. With temperatures 'in the nineties (32–38°C) and humidity to match', the British team won 32–30, assisted by a tank of oxygen and a procession of substitutes, 26 in all. Without them, according to the coach, 'someone would have died out there', only a slight exaggeration in the circumstances.

There have been many other instances of worthy success and heroic failure by British teams and individuals competing against both the elements and their rivals in overseas competitions. Top professional golfers, tennis players, motor racing drivers and athletes have to spend much of their career abroad, following the tours and the major world championships that constitute twenty-first-century sport at the highest level. The conditions they encounter on their travels sometimes bear little relation to those back home, where they first learned their skills. Those who succeed are usually the ones who adapt best to international weather in all its guises.

23

Them Over Here

It may be tough for us, but perhaps we should spare a thought for all those overseas athletes who have the misfortune to play some of their sport in Britain. In recent years, their numbers have increased significantly as a result of the 'shrinkage/growth' principle – shrinkage of the world from separate continents and cultures towards a global village; growth in freedom of contracts and the amount of, and financial inducements in, professional sport. No longer are international visitors restricted to touring rugby and cricket teams, once-a-year entrants to major British events such as Wimbledon and The Open, and European football teams playing in UEFA competitions. There are also large numbers of footballers, and some rugby players, contracted to British clubs; lucrative athletics events in the professional era of the sport; more golf tournaments, more 'World Cups' in cricket and the rugby codes, more and more sport. How do overseas athletes, especially non-Europeans, cope with the fickle British climate?

Although foreign footballers can regularly be seen sporting gloves on the pitch (and who knows whether thermal underwear lurks unseen beneath the team colours), at least they only have to brave the elements for less than two hours at a stretch. The greatest sympathy should still be reserved for golfers and cricketers who have to spend hours and days battling with the weather. But, you are probably thinking, they play in summer. So?

It didn't feel like summer when the Indian cricket team arrived for their tour in 1967. May was wet and cold and the Indian batsmen looked distinctly unhappy on sodden English pitches. It turned out to be the wettest May for nearly 200 years – so wet that Lancashire's match with the tourists scheduled for Old Trafford was transferred to Southport a full week before it was due to be played because of the wet state of the Manchester ground. Fifteen years later it was no better. When India played in a one-day international at Headingley in June 1982, a violent hailstorm the previous afternoon left cars parked at the ground up to their hubcaps in water.

The Indians pleaded in vain that the pitch was unfit for play – the covers had leaked during the storm – and lost by nine wickets.

The 1970s also produced some dismal summers. Cricket writer Alan Gibson thought that Australia were severely disadvantaged in 1977 by their lack of experience of wet English pitches. A team which included Greg Chappell, Rodney Marsh and Jeff Thomson lost the Test series 3–0. The following year it was the turn of New Zealand and Pakistan to suffer. In May, the same writer declared the whole of England and Wales to be 'one vast puddle' and by September he reckoned that the miserable summer had given the visitors little chance to find their best form. Both lost heavily – it makes you wonder why any team would take on the English and their climate!

In 1970 the Commonwealth Games took place in Edinburgh – in July, not necessarily a wise choice, as the city receives more rain in high summer than at most other times of the year. The wind, of course, blows all the time and when the two combine it can feel chilly. The African competitors certainly thought so. Alice Annin of Ghana was second in the women's 100 metres, beaten by 0.1 seconds, but she might easily have won in better weather conditions. She was so cold that she had to be massaged for cramp just before the final. John Sherwood of England won gold in the 400-metre hurdles, beating Kenyan and Ugandan runners into second and third place. After he had battled down the home straight he claimed that the wind was the strongest he had ever experienced and that it had broken up his stride pattern between hurdles. There is no record of what the African hurdlers said! Wind and rain led to 19 of the 43 starters dropping out of a gruelling cycle road race and over-zealous officials disqualified two sprint cyclists when they refused to race on a wet track at the velodrome because they considered it too dangerous. They were later reinstated by the international appeal jury – presumably they too were smitten by the weather and were more sympathetic!

For consistently erratic and frequently dire weather the Dunhill Cup, played at St Andrews in October, probably takes some beating. (And in case you wonder, we're not biased against Scotland – we live here!) Sixteen nations took part in the contest from 1985 to 2000 until the tournament format was changed, normally including teams from the four home nations, Australia, the United States, Spain and Japan. Others have come from more exotic lands – Taiwan, Zambia, Mexico, Thailand and Paraguay – and have struggled to cope with conditions. Sometimes *everyone* has struggled to cope with conditions! The first five years were reasonably pleasant, with some glorious autumn sunshine, interspersed with fresh chilly winds. The captain of the Zambian team, competing in 1986, reckoned that if it got too cold, the team would be in trouble; if they put on too many sweaters, they might not be able to swing! It was just as well they were absent in 1990 – it wasn't only sweaters which were required that year but bobble hats and gloves. The Mexican team, playing in temperatures 20°C lower than those at home, were chilled to the bone. Their captain, Carlos Espinosa, might have won his match against Sam Torrance until he missed a shortish putt at the 17th; he said his hands were too cold.

Thick mist disrupted the event in 1991, but for the next four years it was bitter weather that caused problems, particularly for the overseas teams. When Sandy Lyle, dressed in two cashmere sweaters and waterproofs, commented before the tournament in 1992 that the weather would be a big factor, he was certainly right! The team from Thailand appeared on the links with woolly hats and scarves wrapped around the lower halves of their faces, looking more like escapees from a polar expedition than professional golfers. When they took on the team from America, they scored 80, 82 and 85. Not surprisingly, all the home teams won through and England beat Scotland in the final after four days of wild weather. In 1993 it verged on the unplayable, with temperatures barely above freezing, and icy winds. Fred Couples was seen jogging up and down the fairways between shots to keep warm. In 1994, Chen Tze-Chung of Taiwan, togged out in the inevitable woolly hat, scored 83, the highest winning round in the history of the tournament. The following year, when a team from Fiji played, they allegedly prepared for the Scottish climate by spending time in giant freezers! Sylvester Stallone withdrew from the pro-am competition which preceded the main event because it was too cold for golf. Sometimes even the pros complained. After a particularly gruelling test of stamina in 1998 when howling wind and driving rain led to conditions described politely as 'trying', Ernie Els of South Africa commented that if it was like that back home, golfers stayed indoors!

Considering all the foul, unfamiliar weather, overseas players have coped remarkably well. On an infamous day in 1993, the unfancied and unacclimatised Paraguayans even beat the local Scottish team. In 15 years, the home nations have only got their hands on the trophy four times, although other winners have included Canada and Sweden, no doubt accustomed to wintry golf when the calendar says October. The United States, Australia and South Africa have won most often, leading us to suspect that perhaps the British climate is not as bad as we think. Read on . . .

24

It Also Happens Elsewhere!

Even the most casual trawl through the annals of great sporting nations, such as Australia and the United States, reveals that they too can suffer from lousy weather. As well as rain, fog, wind, hail and snow, other parts of the world have to contend with more extreme and violent climates than that of the British Isles, and weather phenomena such as tornadoes, hurricanes and all manner of storms – dust storms, snowstorms, tropical rainstorms and thunderstorms of great intensity. One of these even managed to extinguish the Olympic flame at the Montreal games of 1976. Many non–European countries also suffer from fierce summer heat which, combined with strong sunlight and sometimes high humidity, can affect both spectator comfort and sporting performance.

These conditions can certainly be expected to cause problems for athletes from more temperate zones. It isn't just Brits, however, who boil, bake and generally overheat, not to mention underperform, in the hot spots of the world – and at least we have the excuse that we're unaccustomed to it at home! In Australia, sometimes even the locals find conditions too extreme. The opening ceremony of the British Empire and Commonwealth Games in Perth, Western Australia, took place on 21 November 1962 in temperatures as high as 41°C, and over 300 people were treated for heat exhaustion. On the hottest November day in Perth for 20 years, the army cadet carrying the nameplate for the Bahamas fainted and had to be carried off on a stretcher, while athletes standing in the centre of the stadium had to be supplied with drinking water as the pageant unfolded. For some of them, this was only the start as the next three days continued to see temperatures of over 38°C. When 18 runners lined up for the 6–mile race (the equivalent of the 10,000 metres) it was windless and stiflingly hot on the floor of the arena: only 7 finished. Surprisingly, Ron Clarke, the Australian world record holder, was not amongst them while the winner, Bruce Kidd of Canada, only survived the ordeal by wearing a baseball cap and a handkerchief round his neck to protect himself from the searing heat. Although the weather then cooled to below 30°C, conditions were still so severe that members of the press corps were predicting fatalities in the marathon, scheduled for mid-afternoon. 'Risk of marathon tragedy – race should be delayed till 5 p.m.,' thundered *The Times*. Officials pleaded for a later start, with those from temperate countries arguing that their athletes could be seriously affected by the sweltering afternoon sun. No one wanted to see a repeat of the Jim Peters incident eight years previously. As it turned out, temperatures dropped with the arrival of rain and the race was won by the European champion, Brian Kilby of England, in conditions more reminiscent of home.

During the first few days of the Games, spectators suffered too. Hundreds watched the rowing events, not from the banks but from the waters of the Canning River. When the rows of metal seats provided for onlookers apparently reached 69°C, many took to the shallows, while some decided that immersion up to the neck was preferable to broiling on the surface. The Australian rowing eight apparently agreed. They held off the New Zealand crew to win by just 5 centimetres in a time of 5 minutes 53.4 seconds, a world record which lasted until 1980, and a remarkable achievement in 33-degree temperatures. Then they promptly jumped into the water to escape the heat – or maybe it was just over-exuberance!

The Danish golfer, Thomas Bjorn, would probably like to have done the same in January 2000. Competing in the Heineken Classic event at Perth in temperatures of 47°C, he described conditions as the fiercest he had ever encountered on the professional golf circuit. Under the circumstances, it was amazing that the top three, Michael Campbell of New Zealand, Bjorn and Alastair Forsyth of Scotland, winner of the European tour qualifying school two months earlier, all hailed from cool climates.

Perth is not the only Australian city to sizzle in the sun, and not all sportspersons conquer excessive heat as successfully as golfers. When Australia played the West Indies on the opening day of the fifth Test at the Sydney Cricket Ground in January 1952, the pitch was perfect, but the temperature was 41°C in the shade and the two teams contrived to lose 19 wickets between them by close of play. The home side was dismissed for 116 while the visitors struggled to 64 for 9, ending their innings on 78, their lowest-ever total against Australia. No explanation could be found for the batting collapses, and the day was described by one historian of Australian cricket as 'bizarre'. Maybe on this occasion the heat had a detrimental effect on the athletes, few of whom would have subjected themselves to the fitness and dietary programmes of today. Even when Dean Jones scored 210 in the first Test at Madras in 1986, contributing to Australia's highest score against India, more rigorous fitness regimes could not save him from hospitalisation. Totally exhausted by his long innings in intense heat and humidity, he required saline treatment for dehydration.

Australia's major cities can all suffer extreme temperatures. By the time the Commonwealth Games took place in Brisbane in September 1982, the marathon was considerately scheduled for 6 a.m. to avoid the sub-tropical heat of the day.

(This is the equivalent of March in the northern hemisphere. Is scorching heat the real reason why Aussie Rules footballers play their winter game in the kind of skimpy vests that Europeans and Americans associate with indoor sports like basketball?) When the same city held the Australian women's hard court tennis championships in 1990, temperatures reached 50°C and Isabelle Demongeot of France, on her first visit Down Under, protested that conditions were ridiculous and she was not coming back! It was even too much for Rachel McQuillan, a native of Oz. While spectators huddled in the shade of the scoreboard and cowered under umbrellas, she wore a hat for protection and poured cold water on her feet at every changeover, but she still developed blisters. The same summer saw ball boys faint at a tennis tournament in Adelaide and players object that the weather was life threatening. A ruling was also made during the 1990s concerning the retractable roof at the Flinders Park stadium, Melbourne. Although it was primarily intended to keep off the rain, the tournament referee for the Australian Open tennis championships in January can order it to be closed if the local weather bureau predicts a top temperature of 35°C or more in the shade. Unfortunately for Martina Hingis, it remained open for her women's final against Jennifer Capriati in 2002.

MARTINA IN MELBOURNE

In her first Grand Slam event after ankle surgery, Martina Hingis was beaten by nerves and the weather. On the hottest day of the Australian Open, she streaked away to a 6–4 4–0 lead against Jennifer Capriati, the title holder, but somehow, at 5–3 in the second set, she failed to convert the first of four match points. In the gruelling tie-break that followed, both players drank water and retreated to shaded areas between points to escape air temperatures of 38°C and a court surface temperature of 46°C. When Capriati won the set 7–6 (9–7), tournament officials allowed the women to take advantage of a rule which affords a ten-minute break between sets. Both players departed for the locker-room to be ice-cooled before the final showdown. Back on court, it soon became obvious that Hingis was nearing exhaustion. She leant on the scoreboard, rested in the shade of the players' tunnel and even lay down at one change of ends. Tournament officials became so concerned that, allegedly, a wheelchair and a saline drip were organised behind the scenes. Finally Capriati took the third set 6–2, the first player to win a Grand Slam final after being match point down since Margaret Court in 1962, and the ordeal was over. In spite of the debilitating conditions, both women spoke to the crowd and the press afterwards, agreeing that it was hard to breathe in the sweltering heat. Hingis said her body was covered in goose bumps – 'it was just dehydration'. Capriati said she had never played in heat like that. It was not a great match in terms of quality tennis but, as drama, it was riveting.

Not all parts of Australasia, however, are renowned for heat. Twelve years after the furnace-like start to the games in Perth, athletes shivered in temperatures of 11°C at the opening of the Commonwealth jamboree in Christchurch, New Zealand. The Australian team manager was heard to remark that the ceremonies were 'too long and too cold. I hope to God it hasn't affected anyone on the team.' The final medal tally suggests it hadn't – Australia topped the table with a total of 83 medals, two more than England. Cricket in New Zealand can also be a miserable affair, as the England touring team discovered in 2002. But 'twas ever so! Eden Park, Auckland, was the scene of a draw between England and the home side in March 1959, in which the first two days saw blustery weather that frequently blew the bails away and the last two were completely rained off. Although the first Test against South Africa at Basin Reserve, Wellington, in February 1964 is likely to be remembered for the pitch damage caused by anti-apartheid demonstrators, the unseasonably cold weather was commented on in *Wisden*. It failed to improve for the second Test in Dunedin. In cold, drizzly rain, only four hours of play were possible on the first two days and the match resulted, almost inevitably, in a draw. As for the second Test between New Zealand and England in 2002, not only was the first day at Basin Reserve washed out but storm-force winds resulted in the covers being anchored to the ground with boulders.

Old Trafford may have the worst reputation in cricket for 'rain stopped play', but the Wellington ground probably runs it a close second. It has hosted far fewer Tests than its English counterpart but its post-war weather statistics make for gloomy reading, with the 1980s providing the soggiest scenarios. Four out of nine Tests lost entire days to rain. The matches against Australia in 1982 and England in 1988 lost two each and the 1986 Test against the Aussies was little better with only 81 minutes possible on the fourth day and none on the fifth. The resulting sequence of draws, unbroken from 1984–89, is far worse than the comparable figures for Old Trafford where, amongst the inevitable stalemates, there were three victories, all for the tourists. Perhaps this says as much about English cricket as about the English climate!

Wellington is hardly an advert for New Zealand weather, but it is possibly just as well that only a handful of Tests have been held at Dunedin. Apart from two rain-affected drawn matches in the 1960s, it has the dubious distinction of playing host to one of only five Tests since 1877 to be abandoned without a ball being bowled, against Pakistan in 1989. Another, already mentioned in the previous chapter, was at Georgetown, Guyana, and three involved England and Australia – so no marks for guessing that one of the other venues was Old Trafford, where there were two total wash-outs in 1890 and 1938. Australians will not be surprised to learn that the other culprit was Melbourne, one of the few Australian cities to experience four distinct seasons, sometimes – as non-Melburnians will gleefully tell you – all in one day! In 1993, when the city was suffering some of its wettest and coldest December weather since records began 150 years earlier, the Melbourne Tennis Classic had to be moved indoors, a 40-knot gale resulted in the postponement of two major boat races, and the historic first Test to celebrate the return of the South African cricket team to Australian soil after an absence of 30 years was virtually washed out. No play was possible on the second day and only two hours on the third. Dr Ali Bacher, managing director of the United Cricket Board of South Africa, was philosophical. After such a long wait, he observed, what were a few more days?

In any case, citizens of other state capitals shouldn't laugh at Melbourne, as Australia's overall reputation for hot, dry summers and a relatively benign year-round climate cannot stand up to close scrutiny. When Adelaide was home to the Australian Grand Prix between 1985 and 1995, it became notorious for its rainstorms. In November 1989, Thierry Boutsen of Belgium won a race which had originally been halted after two laps and which some, including Alain Prost, felt should never have been restarted. He declined to line up a second time. In 1991, with the constructors' championship still to be decided between Williams and McLaren, the race was abandoned in torrential rain after only 14 laps, making it the shortest ever Grand Prix event. Ayrton Senna, already world champion, was leading at the time, ensuring that the constructors' title went to McLaren. The authorities came in for some criticism when the rain stopped later in the afternoon, but blamed the information they had received from the weather officer at Adelaide airport, who had forecast a dry start to the race and rain later instead of the other way round. Adelaide finally lost its flagship motor race in 1995 – to Melbourne!

Brisbane's cricket ground, The Gabba (officially Woolloongabba), is another sports venue regularly affected by violent thunderstorms, probably none more dramatic than in 1946 when E.W. Swanton, the English commentator, reported hailstones the size of golf balls and the sight of the stumps and tarpaulin wicket covers floating briskly downwind on the flood waters, only to be halted by the boundary fence. The second of two storms to affect the first Test against England, it did nothing to help the visitors' chances. After the deluge and the 130-kph gale had ceased, play resumed but England lost 15 wickets in three and a half hours, handing Australia a record victory, for them, of an innings and 332 runs. At the same venue, Pakistan were saved from a second successive innings defeat in 1983. Torrential rain flooded the ground after lunch on the fourth day and washed out the remainder of a match which had seen Greg Chappell record his twenty-third Test hundred and Allan Border his tenth.

The seas around Australia, but particularly off the south coast, are also well known for their treacherous conditions, even in summer. When 115 yachts set out from Sydney Harbour on Boxing Day 1998 for the 54th Sydney to Hobart race, over a distance of some 1,000 kilometres, they knew that ahead lay the notorious Bass Strait between Tasmania and the mainland. The following day, the fleet was hit by 10-metre waves and 145-kph winds. Four men died, two others were swept overboard, presumed drowned, and 56 crew members were plucked from the sea. Nearly two-thirds of the yachts retired and an inquiry was launched into the worst tragedy ever to hit the race. Although increased safety measures have now been introduced, the most significant consequence of the disaster has been a big drop in the number of competitors. Higher insurance premiums have made the race much more expensive, and in 2001 there were only 75 entries.

They also faced severe weather. A number of yachts found themselves in the path of a waterspout, the sea equivalent of a tornado, with no means of taking evasive action. Others reported massive hailstones. Although there was some damage, there were no casualties, but the weather still dealt a blow to the chances of the leading yacht, *Nicorette*. Her mainsail was ripped in the hailstorm and had to be replaced, and she dropped back to fourth place. The following day when, perversely, the crews had to contend with a complete absence of wind, *Nicorette* put in a late challenge and eventually finished second, only 20 minutes behind the winner. Don't tell any yachtsman that Australia is a sunny land of beach, barbecue and beer. It also has its dark side.

An even greater range of severe climatic problems affects North America, where bitter cold and heavy snow contrast with extremes of heat and humidity. Violent winds and storms are also common and golf is particularly susceptible to these conditions. The American Professional Golfers Association tour is said to follow the weather, staying south during the early and late parts of the season and only venturing north from May onwards, but every year sees tournaments suspended, shortened and significantly affected by thunderstorms, heavy downpours and lightning strikes. With almost half of the tour events taking place in the central states and Florida, the areas with the highest storm frequency, this is hardly surprising. One of the most widely reported incidents involved Lee Trevino and his two playing companions, Jerry Heard and Bobby Nichols. They were struck by lightning while playing the second round of the Western Open at Oak Brook, Illinois in 1975 and ended up in hospital where they were all said to be 'a little shook up and sore'. Trevino suffered burns to his back and left shoulder, and was advised to rest for a few weeks. The following year, when the US PGA tournament took place at Bethesda, Maryland, the whole event was played in muggy, sweltering heat, the third round was interrupted by heavy rain and the final 18 holes were postponed because of thunderstorms. The entire field, including the 19 players who had already completed rounds, had to play again the next day, thereby preventing several Americans from flying to Scotland to compete in a tournament at Gleneagles. In 1983, a severe storm delayed the climax of the US Open at Oakmont, Pennsylvania, when Tom Watson and Larry Nelson were level with three holes to play. They resumed on Monday morning and Nelson won by a stroke. More recently, Tiger Woods won the Memorial Tournament at Dublin, Ohio, in May 2000 after lightning and torrential rain had halted the final round, forcing players to return for an extra day, while the following year the Ladies Professional Golf Association Champions Classic at Beavercreek was shortened to 36 holes for the first time in its history after a series of storms.

American heat and humidity have influenced numerous sports events. When Sugar Ray Robinson challenged Joey Maxim for the world light-heavyweight championship at the Yankee Stadium, New York, in June 1952, he was beaten as much by the heat as by his opponent. In a temperature of 40°C, the referee collapsed at the end of the tenth round and had to be replaced! Meanwhile Robinson, well ahead on points, was so dehydrated and exhausted that he was unable to come out for the 14th round. When Ken Venturi won the US Open in 1964, temperatures were as high as 35°C with relative humidity of 80 per cent. At a time when the two final rounds were played in one day, he was so dehydrated that it was feared he would be unable to compete in the afternoon. He was accompanied by a doctor over the last 18 holes and somehow managed to fight off his exhaustion to win by 4 shots. When the Formula One Grand Prix circus descended on Dallas in 1984, the track broke up in the fierce heat during practice and had to be patched overnight with concrete. When this proved to be unsuccessful, the scheduled warm-up had to be cancelled to allow further repairs with one-hour-setting cement. The race itself was described as one of the most exhausting for many years. There were several crashes, Nigel Mansell collapsed while trying to push his broken-down car over the line and Dallas never held the event again. And when the United States was chosen as the venue for the 1994 World Cup and Atlanta, Georgia, as the site for the 1996 Olympics, the controversies – of which there were many – included the impact of hot and humid American summers on competitors.

Spare a thought for the American fans too. At outdoor boxing venues, they faced the possibility of sunstroke or saturation. The day Jack Dempsey beat Jess Willard for the heavyweight championship in July 1919, a crowd estimated at 150,000 sweltered in the heat of Toledo, Ohio, and hundreds fainted. Seven years later, those who witnessed Dempsey's defeat by Gene Tunney in Philadelphia were soaked to the skin in driving rain. Worse still, spectators risk death on the golf course – lightning fatalities occurred at both the US Open at Hazeltine, Minnesota, in 1988 and the US PGA tournament at Crooked Stick, Illinois, in 1991. They can also suffer frostbite at the football stadium. When the San Francisco 49ers met the Chicago Bears at Soldier Field, Chicago, in the season 1988–89, the temperature dropped to minus 12°C with a wind chill of minus 32°C and fans were warned, 'If you experience tingling, numbness or change of skin colour, seek help.'

They remain stoical, however. American football in the northern states during December and January regularly takes place in snowstorms, which must be tough on devotees of the game but even more difficult for those cosseted southern teams, like the Houston Oilers, who play their home matches indoors on artificial grass. Even when a sport is held under cover, spectators still have to get there. The doggedness of the 15,000 who ploughed through thick snow to watch Jack Kramer's first professional tennis match at Madison Square Garden, New York, certainly impressed him. Having won Wimbledon and the US Open in 1947, he was determined to make money from the game. He may have lost that first pro match against Bobby Riggs, but the fans paid $250,000 to watch him do it on a wintry January night, convincing Kramer that the professional tour was a money winner. By 1951 he had taken control of the international professional tennis circuit, and he went on to sign major championship winners such as Ken Rosewall, Pancho Gonzales, Lew Hoad and Rod Laver.

If there's one type of weather America does well, it's snow and blizzards. The world's greatest recorded snowfalls have all taken place in the United States and the 'Great American Winter' of 1978, when the north-eastern states saw exceptional snowstorms, makes the British winters of 1947 and 1963 look puny by comparison. Remote spots in Wales and the Pennines may have registered accumulations of up to 150 cm in 1947, but the American city of Buffalo, the second largest in New York state, received three times that amount in the winter of 1978. No problem, you would think, if the Winter Olympics are held in North America. But when Lake Placid, just over 400 kilometres from Buffalo and the same distance upstate from New York city, staged the event for the first time in February 1932, snow had to be imported from Canada in large trucks. A sudden thaw also caused the four-man bobsleigh to be held *after* the official closing ceremony. Lake Placid also hosted the Games in 1980 and this time wind was an added problem. Not only did the cross-country ski course have to be packed with artificial snow to improve the depth and surface condition, but gusts of up to 120 kph jeopardised several races, including the women's Alpine downhill. Part way through the 70-metre ski-jump, conditions were deemed so dangerous that, although nine skiers had already jumped, the starting point was moved lower down and everyone had to start again.

It was also wind that brought havoc to the schedules of the Winter Olympics in Calgary, Alberta, in February 1988. The chinook, a warm dry wind which affects the eastern Rockies, suddenly blew into town, raising temperatures from minus 23°C at the opening ceremony to plus 6°C. It disrupted the ski-jumping and dumped so much dirt and grit on the bobsleigh course that, according to some crews, the normally smooth surface felt like sandpaper. The spring-like warmth melted snows and exposed the ski-jumps, making them potentially dangerous, while the Nordic combination event, comprising ski-jumping and cross-country skiing, had to be contested all in one day in case the snow disappeared completely.

The most recent Games, at Salt Lake City, Utah, in 2002, did not entirely escape without problem weather, particularly in the first week. Even before the official opening, training runs for the downhill skiing and qualifying rounds for the ski-jumping had to be rescheduled because of blizzard conditions, with winds gusting to 95 kph. The women's downhill competition suffered several delays and postponement as high winds and sometimes poor visibility continued to affect the mountain venues. It was snowing when Alex Coomber, the British competitor in the women's skeleton, set off on her second run, but the weather did not affect her final bronze medal position. At least the British gold medal winners, the women's curling rink captained by Rhona Martin, did not have to contend with the elements. Curling has been largely an indoor sport for nearly a century.

Every continent has its winter weather problems. Deep snow and sub-zero temperatures are essential if skiing and snowboarding are to take place, but relatively dry and still conditions are needed on competition days. Ask the organisers of the 1998 Winter Olympics at Nagano, Japan, about heavy snowfalls. The opening days of the games were hampered by blizzards, with the entire Alpine skiing programme having to be rescheduled several times. The men's snowboarding giant slalom, in its debut as an Olympic event, began in bright sunshine but the second run was completed in poor visibility, resulting in much slower times for the competitors, several of whom lost their way in the snow. As the blizzards continued, troops were despatched to clear the downhill and slalom ski runs, a task which required round-the-clock shifts using floodlights. With five days of Alpine competition lost to the elements, the organisers finally had to arrange three events for the same day, an unprecedented situation for the Winter Games, although equally poor conditions at the 1993 World Ski Championships in Japan meant that one major race was never completed. Astonishingly, none of this was unexpected as the International Ski Federation knew from meteorological records that they were unlikely to get more than four days of suitable downhill weather during the Olympic fortnight. Nagano apparently lies in the path of cold, dry winds which cross the sea from Siberia, picking up moisture ready to deposit on the nearest land mass – Japan!

Most sports have been affected by weather, sometime, somewhere. The glare of the sun during the 1962 World Cup final between Brazil and Czechoslovakia in Santiago, Chile, was held responsible for the goal that sealed victory. The Czech goalkeeper, Schroiff, was momentarily dazzled and dropped a high ball. Brazil scored their third goal from this error and won 3–1. The South African golfer, Darren Fischardt, won his maiden European tour title in the Sao Paulo Brazil Open in March 2001 after lightning during the first three days reduced the event to 54 holes. A World Cup hockey semi-final between Malaysia and India in Kuala Lumpur was abandoned in 1975 after the pitch became waterlogged in torrential rain. The Mexico Olympics of 1968, already feared by competitors because of the altitude and heat, added rain to the list of potential hazards. The cross-country course for the three-day event was completely flooded in places and riders were unable to see some of the take-off points for fences. On the other hand, a shower of rain on the day of the long jump final removed the worst of the afternoon heat, ensuring perfect conditions when Bob Beaman broke the world record, one which stood until 1991. Even the desert state of Dubai is not immune from the elements. In March 1997, the richest horse-race in the world, the Dubai World Cup, had to be postponed after the heavens opened, delivering nearly half of the country's annual rainfall in one afternoon. It was three days before the racetrack at Nad al Sheba could be reopened.

Cricket has been rained off wherever the game is played. Pakistan suffered a particularly sodden series against the West Indies in 1980–81 when three of the four Tests lost a whole day's play to rain. At Lahore and Karachi it was the first time this had happened for 15 and 17 Tests respectively. When Brian Lara amassed his record Test score of 375 in the fifth match between the West Indies and England in Antigua, he kept his concentration in spite of four rain breaks on the second day. Sri Lanka were undoubtedly hampered by bad weather in the quadrangular one-day Mandela Trophy, held in South Africa in season 1994–95. Two of their four qualifying matches were rain affected: the first, against New Zealand, resulting in a flooded ground at Bloemfontein; the second against South Africa at Port Elizabeth, reduced to 34 overs. Over-enthusiastic spectators spent part of the two-hour rain break sliding across the tarpaulin covers!

Europe has also had its share of awkward sporting weather. Snowy pitches, foggy football stadiums, golf course thunderstorms and rain-affected Grand Prix circuits are not uncommon closer to home, and although the frequency and intensity of these climatic episodes tend to be fewer and less dramatic than their counterparts in Australia, Asia and the Americas, they are no less disruptive. The European Cup quarter-final in 1987 between Besiktas of Turkey and Dynamo Kiev was postponed three times because of snow, and eventually switched from Istanbul to Izmir. Cardiff City's 1968 second-leg quarter-final in the Cup-Winners' Cup against Moscow Torpedo had to be played in Tashkent because of the severe Moscow winter. Supporters of Dukla Prague will not need to be reminded that their third-round second-leg UEFA cup tie against Inter Milan fell foul of the well-known San Siro fog twice in December 1986. Fewer are likely to recollect that the England Under-21 game against Italy in November 2000 was abandoned after only 11 minutes because of fog in the Monza stadium. Golf fans, however, will never forget the scenes at the beginning and end of the Ryder Cup at Valderrama in southern Spain in 1997 when storms and torrential rain threatened to wreck the tournament. Occasional heat waves in Europe have also made life difficult for golfers. The first round of the Scandinavian Masters in 1994 took place on the hottest July day in Stockholm for 100 years, with temperatures in the 30s. It was not a pleasant day for golf. Historians of the Olympics may recall that the torrid heat of the Paris games in summer 1924 was even more unpleasant for athletes. When the 10,000 metres race took place in temperatures of over 40°C, only 15 of the 39 runners completed the course.

Not surprisingly, however, instances of cold, wet, windy or snowy weather affecting European sport are more commonplace than examples of blistering heat. The Parisian winter of the same year was wet enough for the Seine to flood the rugby pitch at Colombes, where Scotland were scheduled to play France. (This had also happened in 1913). A French report of the same fixture in 1926 told of heavy rain and a ball which frequently disappeared into the waterlogged ditch around the playing area. The groundsman had to fish it out with a butterfly net, prompting the comment, 'Il faut bien rire un peu.' ('You had to laugh!') Grand Prix motor racing has frequently been hampered by heavy rain, with the circuits at Spa in Belgium and Hockenheim in Germany particularly notorious for their inclement weather, each with its own microclimate. The European figure skating championships, held at an open-air stadium in Budapest in 1963, were interrupted by falls of snow and sleet, and it was no surprise when the International Skating

Union ruled in 1967 that all future international championships had to be staged indoors to eliminate the lottery of the elements. However, just as the severe winters of 1963 and 1979 allowed curling's Grand Match to take place in Scotland (see Chapter 9) so the harsh January of 1997 enabled the Dutch to hold the legendary Friesland Eleven Cities skating marathon for the first time in 11 years along a 200-kilometre route of frozen canals. There were nearly 16,000 participants, but the extreme cold and gruelling distance led to hundreds being treated for fractures, frostbite and exhaustion. Perhaps it's just as well it only happens infrequently.

Overall, the worst sporting weather occurs outside Europe. A final example, from one of those spectacular round-the-world races whose *raison d'être* seems to be gaining maximum publicity for their sponsor, might prove the point. One of the earliest of these extravaganzas was the *Daily Mirror* World Cup Rally. It started in April 1970 from Wembley Stadium, scene of England's football triumph in 1966, and finished at the Aztec Stadium in Mexico City, soon to stage the next final. Ninety-six teams were sent on their way by Sir Alf Ramsey, the England manager, one of them including a former member of his squad, Jimmy Greaves. The first part of the race followed a 7,240-kilometre route from England to Bulgaria, then back to Portugal, visiting 11 European countries. The competitors then boarded ship for South America, landing at Rio de Janeiro, and headed south to Argentina. Crossing into Chile, they drove north via Bolivia, Peru, Ecuador, Columbia and Central America to Mexico, a total distance of 25,750 kilometres.

The weather they encountered was certainly varied. A bright spring morning in London became a wet night across the Channel. There was deep snow and diversions in Yugoslavia and mist patches in Portugal, but the 25 retirements notched up on the European leg were all the result of mechanical failure and accidents. Conditions were more extreme in South America. Tropical downpours left rivers in spate, bridges almost under water and dirt roads turned to mud. Casualties mounted. Unexpected rains in Argentina changed the dry Pampas to a glutinous quagmire and cars became bogged down. It continued to rain in Chile, while snow in the high Andes blocked the route, causing more detours. Only 39 cars made it to La Paz, the capital of Bolivia. Altitude problems in Peru gave way to heat and humidity in Columbia, where the climate was described as 'hot and horrible', and 'a kind of open-air Turkish bath'. After three weeks of driving, 23 crews finally made it to Mexico, only a quarter of the original field. Jimmy Greaves, who finished sixth, commented, 'It's tougher than soccer.'

One thing is certain. Britain does not have the worst sporting weather in the world – the most unpredictable, perhaps, the most talked about, maybe, but not the most extreme.

SATURATION POINT

25

And if you Thought Things had Improved –
A Nine-Month Diary of Sport and the Weather

2001
1 August
Umpire Jeff Evans, officiating in his maiden season of first-class cricket, gave a Leicestershire batsman not out when most other observers – especially the Essex team – thought he clearly was. A few minutes later Evans collapsed at square leg with heat stroke, a result of standing all day without a hat.

2 August
Rain stopped play at Canterbury and Colwyn Bay in the cricket County Championship and the one-day game between Warwickshire and Nottinghamshire was abandoned.

3 August
Rain stopped play again at Colwyn Bay.

4 August
Rain finally ruined the match at Colwyn Bay between Glamorgan and Lancashire.
   Play was suspended at the Weetabix Women's Open Golf at Sunningdale because of rain, though several players felt the decision came too late, as they were forced to putt on greens so wet that they had to be squeegeed.

7 August
No play was possible on the first day of the Roses cricket match between Yorkshire and Lancashire at Old Trafford. This 12th wash-out of their season did not help Lancashire, who were defeated on the final day with over two hours to spare.
   Thirty-knot winds at Cowes meant that 629 of the 1,005 entrants, mainly the smaller boats, remained at their moorings.

8 August

There was a full day's play at Old Trafford, though every other county match was either abandoned, did not start, or, at best, was interrupted.

9 August

Only six golfers actually hit a ball on the first morning of the Wales Open at Celtic Manor before play was suspended because of heavy rain. They included David Dixon, making his professional debut, who scored a bogey at the first hole. The rain continued, the course became waterlogged and play was later abandoned for the day.

12 August

One-day cricket became two-day as the semi-final of the Cheltenham and Gloucestershire tournament between Lancashire and eventual winners Leicestershire was forced into a reserve day because of rain. Elsewhere, rain interrupted play for an hour in the Second Division one-day cricket match between Durham and Derbyshire at the Riverside. The game should have resumed at 7.10 p.m. with Derbyshire needing 56 runs off only 13 balls. However, at 7.02 p.m. a few more spots of rain signalled the abandonment of the match and, under the Duckworth/Lewis system, Durham were declared the winners by 33 runs.

The Wales Open squelched to a climax. Thirty-six holes were abandoned and the three Saturday night leaders played-off on the par-three hole which did not require use of the waterlogged fairways. After five attempts, Irishman Paul McGinley triumphed over Englishman Darren Lee and Scot Paul Lawrie.

Tim Henman lost to Brazilian Gustavo Kuerten in the semi-final of the Masters Series in Cincinnati. The match took place during colossal evening storms in which huge zigzags of lightning struck the site and eventually brought play to a halt.

15 August

The Italian yacht *Stealth* finished ahead of the other 232 starters in the Rolex Fastnet race but failed to beat the course record because of the lack of even a mild breeze in the final stages. The last 16 kilometres of the race took almost 6 hours to complete.

19 August

Poor weather kept the crowd at the IAAF Norwich Union Classic some 2,000 below the capacity of the Gateshead Stadium.

Clouds and rain allowed only 25 overs to be bowled at the Headingley Test match, but the final delivery by Glenn McGrath was worth £200,000 as it saved the ECB from paying a refund to the crowd. One ball fewer and the spectators would have received half their ticket money back.

More than 10 per cent of the fleet pulled out of the America's Cup Jubilee regatta when the Solent turned rough, with a 25-knot south-westerly whipping the sea into boat-breaking waves. The nine modern carbon-fibre America's Cup boats were considered too delicate even to leave their moorings.

22 August
Too little wind at Cowes curtailed most of the racing. Of the 200 Jubilee yachts there to celebrate the 150th anniversary of the America's Cup, only the dozen Grand Prix 12-metre boats and nine America's Cup yachts managed to race.

23 August
Rain interrupted play in county cricket matches at Colchester, Taunton and Northampton.

The opening round of the Scottish PGA Championship at Gleneagles was delayed for three and a half hours by local fog. Golf finally began at 11 a.m., but bad light eventually led to the suspension of play with 26 groupings still on the course.

25 August
Arsène Wenger, the Arsenal manager, described the conditions at Highbury as 'African heat', but his side still beat Leicester 4–0.

29 August
Rain and bad light continually interrupted the penultimate day of the final Under-19 cricket Test match at Chester-le-Street between England and the West Indies.

31 August
A wet track at Spa-Francorchamps led to David Coulthard crashing his car after only three practice laps for the Belgian Grand Prix. Approaching a corner at 230 kph, he lost visibility in the plumes of spray and went into the barriers. The shunt wrecked his car, tearing off the front wing and right front wheel. Fortunately, he was able to walk away and race the following Sunday, gaining second place behind Michael Schumacher, who had also crashed on the same day as Coulthard.

3 September
The Duckworth/Lewis system was used to determine victories for Surrey over Yorkshire and Gloucestershire over Leicestershire, the latter ultimately a fatal blow to the division leader's Norwich Union title hopes.

5 September
Cricket again suffered from the weather. A downpour during the lunch interval at The Oval meant that 20 overs were lost on a murky day in which county champions Yorkshire, with only four capped players in the team, struggled to 179 for 5. At the Riverside, 24 overs were lost to morning rain in the match between Durham and Worcestershire.

International rower, Carolyn Jones, was in a critical condition in a Swiss hospital being treated for hypothermia. She became trapped between rocks during a canyoning event in which participants body-surf and climb down fast-running mountain rivers, often through gorges and steep ravines. As she was body-surfing the rapids in Via Mala gorge, she was swept underwater and became clamped between two rocks. Her team-mates and a local mountain rescue team struggled for 50 minutes to free her as she lay unconscious in the cold water before divers arrived to drag her to safety through an underwater cave. Jones won a silver medal in the women's lightweight single sculls at the British National Rowing Championships in 2000 and was Scottish national champion in the event.

7 September
An almighty storm and torrential rain before kick-off in a rugby league match between St Helens and Leeds resulted in a very wet ball, many handling mistakes and dropped passes. In the vicious swirling wind, kicks at goal were being missed because players were unused to holding the ball still for the kicker.

The blustery wind was so strong in the Women's PGA Matchplay Championship at Gleneagles that, on the seventh green, it moved Catrin Nilsmark's ball after she had addressed it. This cost her a penalty stroke and gave her opponent Laura Diaz a half. Ultimately Nilsmark lost by one hole.

9 September
The Duckworth/Lewis method was used to give Durham a nine-run victory over Worcestershire in the Norwich Union League.

10 September
Golfer Nick Faldo is reported to have said that, although the American Ryder Cup team looked stronger on paper than the European side, the match was at the Belfry. He said, 'Hopefully the good old British weather might also play a role. It's early September and already it's started to turn.' (It never got the chance because of the terrorist attacks on the United States the following day which led to the cancellation of the event.)

12 September
No play was possible after the morning session at Scarborough between county cricket champions Yorkshire and bottom-of-the-table Essex.

13 September
Cricket was again disrupted almost everywhere. Only 31 overs could be bowled in four short sessions at Trent Bridge; there were numerous rain hold-ups at Worcester; lashing rain limited play to 18.4 overs at Derby; and north of the border, at Linlithgow, the start of the Cheltenham and Gloucestershire Trophy match between Scotland and Dorset was delayed till 2.45 p.m. and three further

interruptions limited play to 39.1 overs. At Hove, however, it remained relatively dry – only two overs were lost – and Sussex were able to claim seven bonus points to close in on the Second Division title.

15 September

In torrential rain at the Reebok stadium, Premiership leaders Bolton lost their unbeaten record going down 1–0 to previously goalless Southampton. No doubt the rain, swirling wind and a greasy pitch contributed to the reversal of the formbook.

The 112th County Championship ended with a whimper in the wet. Lancashire versus Kent was abandoned because of rain, the 11th total loss of a day's play at Old Trafford that season. Worcestershire gained a draw in a rain-halted match at Derby which was sufficient to bring them promotion to the First Division. Hampshire were foiled in their victory bid at Trent Bridge by a downpour that ended play an hour early. Heavy overnight rain delayed the start at Sophia Gardens, Swansea, until 1 p.m. but Surrey chose to bat on from 572 for 3 to a meaningless 701 for 9, a decision that brought a rebuke from Glamorgan's chairman, who accused them of being unsportsmanlike.

In a storm-force headwind, jockey Mick Kinane rode Milan to a five lengths St Leger victory to give Aidan O'Brien his seventh European Classic win of the season. Kinane kept the horse sheltered in the pack until they reached the four-furlong post.

16 September

Strong winds hampered world half-marathon champion Paul Tergat's attempt to complete the Great North Run in under an hour. He won, but was 30 seconds outside his target.

The one-day cricket league ended with the Duckworth/Lewis system bringing victory to Kent, who were then crowned Norwich Union champions.

18 September

Racing at Yarmouth was abandoned due to waterlogging. There was nearly 14 mm of rain overnight, bringing the total to 43.8 mm over the previous seven days.

19 September

Yarmouth racing was again called off after an early morning inspection.

20 September

Racing at Yarmouth was washed out for the third consecutive day after torrential overnight rain saturated an already waterlogged course. This was the first time that Yarmouth had lost a full three-day meeting and hence they had not insured against the possibility.

21 September

No driving was allowed on the first practice day for the Rockingham 500 race for turbo-charged Champ Cars, because the new Northampton track was suffering from 'weeping' – the seepage of water through the surface due to a raised water table, brought about by incessant rain. The drilling of 50 boreholes saved the race, along with the use of hot air blasted from jet turbines on the backs of lorries.

27 September

MCC members voted to increase subscriptions by 50 per cent, a decision that will allow three key projects to go ahead at Lord's. The extra money will finance a feasibility study of the installation of floodlights which could be used to make up time lost through rain or bad light; the entire outfield will be relaid with a fast-draining top soil; and 'drop in' pitches prepared outside the ground will be introduced for some games.

1 October

The All England Club started work at Wimbledon to install two new clay courts. These have low maintenance costs and can be used virtually all year round. The surface is frost resistant and drains quickly enough to prevent prolonged delays to practice sessions or matches.

4 October

Catherine Ashton, Minister for School Standards and Early Years, explained that the government would support schools that want to sell off a sports field that is constantly waterlogged, provided that the money was used to build an all-weather pitch or changing rooms.

7 October

Paula Radcliffe retained her world half-marathon title in Bristol. Although she improved her European record by 20 seconds, she missed the world record by 4 seconds, a failure attributed to a buffeting wind on the second half of the course.

When rain began to fall at the British Touring Car Championship's final race of the 2001 series, many of the drivers headed for the pits to change tyres, but Anthony Reid, in an MG, took the gamble to stay out on slick tyres. His decision paid off as he was in the lead when the event was called off because of the driving rain and gale-force winds.

10 October

The race meeting at Lingfield Park – not on the all-weather track – was abandoned due to waterlogging. Racing had taken place the previous Friday on ground that was heavy and soft in places, and since then there had been 57 mm of rain.

17 October

Lingfield again suffered from the weather as racing was abandoned halfway through the afternoon, although there had been no rain since the course was passed fit for racing at 8.30 a.m. the previous day. Unfortunately, the starting stalls became bogged down in soft ground and, after a tractor towed them out of the mire, the round course was too churned up. One race was run over the straight six furlongs, using a tape start, but the others could not go ahead.

18 October

The new-format Dunhill Links Championship, a glorified pro-am event with £3.5 million in prizes at stake, had to be suspended as mist enveloped the Kingsbarns course, one of three Fife courses being used for the tournament. Many of the golfers would have been relieved, given the exceedingly chilly, wet and breezy conditions.

19 October

The Fife weather did not relent, and play was once again suspended in the Dunhill Links Championship.

20 October

And again.

Also in Scotland, the Premier League game between Dundee and Motherwell and the rugby fixture between Kirkcaldy and Heriots FP were called off because of waterlogged pitches.

21 October

Once more the players in the Dunhill Links Championship faced horrendous conditions and play was called off after they had finished the delayed third round. The final round was played in cold but drier conditions on the Old Course on Monday, where Scot Paul Lawrie holed a 35–60-foot putt (depending on which newspaper is believed) on the last green to win the first prize of over half a million pounds. This was his first tournament victory since the Open Championship of 1999.

23 October

Yarmouth racing was called off, the fourth consecutive day's racing lost at the Norfolk track because of waterlogging.

24 October

Racing at Nottingham was cancelled because of waterlogging.

26 October

Flood waters at Okehampton, Devon, claimed the life of 12-year-old Charlotte Saunders when she was thrown from her pony while trying to ford the River

Okement. Normally only a few inches deep, rain over the previous two days had swollen the river into a raging torrent.

Heavy rain, which left standing water on some parts of the track at Doncaster, forced the abandonment of two races on the round course.

27 October
The *Racing Post* Trophy at Doncaster was won by High Chaparral over the straight mile, rather than on the nominated round course where the ground was described by the sponsors as 'almost bottomless'. Racing at Newbury was abandoned because of waterlogging.

29 October
Racing was abandoned at Leicester after a Saturday inspection because of waterlogging. There had been 65 mm of rain over the week with 18 mm on Friday alone.

31 October
The winner of the first leg of the Volvo Ocean Race, *Illbruck*, was accused of cheating by making illegal use of an internet weather site.

2 November
Play was suspended due to lightning in the Italian Open in Sardinia and the first round had to be completed on the following morning.

8 November
The English Institute of Sport announced the creation of the £10 million custom-built Redgrave–Pinsent Rowing Lake at Caversham in Berkshire. It was partly a response to the previous year's variable water conditions on rivers, which seriously affected rowing.

11 November
Padraig Harrington won the Volvo Masters in Jerez, Spain. Only three rounds were played after Saturday's round was abandoned in fierce winds. Gusts of up to 80 kph caused balls to move on three of the more exposed greens and persuaded tournament director David Garland that the conditions would be unfair on players. The same thing had happened in 1997 when the tournament was similarly decided over only 54 holes.

13 November
Lingfield's £3 million Polytrack surface came through its first public trial with acclaim. Jockeys liked it because, unlike the former Equitrack surface, there was little kickback. The owners, Arena Leisure, believe that it will be genuinely all-weather, a matter of some importance to them as they had lost many fixtures to the frost the previous winter.

19 November

In India, England's assistant coach, Graham Dilley, was taken to hospital suffering from the effects of heat after working with players all day in the Mumbai sun.

21 November

Leeds United beat Grasshoppers of Zurich 2–1 in the away leg of their UEFA Cup tie in conditions that many referees would not have tolerated. The sodden pitch made the game a lottery, with passes falling short of their target as the ball stopped in the surface water.

24 November

The rugby union match between Wales and Australia at the Millennium Stadium was played with the roof closed because of the wet weather.

At La Manga in southern Spain, finals day in the *Sunday Telegraph* David Lloyd Leisure Grand Prix was washed out by torrential rain. This tournament for young British tennis players had been switched to Spain to give them experience of playing on clay.

1 December

Waterlogged pitches caused the first postponements of the football season. In England, Peterborough v Stoke was called off and in Scotland, Elgin v Dumbarton. It was not a hard day's work for the Pools Panel, who assessed the former fixture as a home win and did not have to consider the Scottish Third Division fixture as it was not on the coupon.

The 11–4 favourite for the Hennessy Cognac Gold Cup, Montifault, was pulled up before the second-last fence as the horse could not cope with the 'rain-sodden' ground.

10 December

In response to complaints by jockeys at Folkestone that they were being blinded by the low sun, the clerk of the course had two fences dolled off to prevent them being jumped. Punters then complained that a dodgy jumper, Elegant Spirit, had been helped to victory in the novice chase by having to face only 11 fences instead of 15 on the two circuits of the track.

11 December

The Scottish Cup was hit by wintry weather as the second round replay between East Fife and Queen's Park was ruled out following a referee's lunchtime inspection that revealed a number of icy patches on the pitch.

Sedgefield racing was called off because of the frosty ground.

14 December

Friday night's Second Division match between Brighton and Reading at the

Withdean Stadium was frozen off. In Scotland, referees were put on standby to make Friday inspections if clubs had any doubts so that fans could get an early warning.

15 December

All four scheduled National Hunt meetings had precautionary early morning inspections. Two failed. The richest jumping card since April was abandoned when Cheltenham's Prestbury Park course was frost bound. Two of the major races were transferred to Newbury's Wednesday meeting and the three-mile novice hurdle to Warwick for the following Saturday. Freezing weather also claimed Lingfield.

16 December

In rugby union, the Mid-Wales League programme was wiped out by icy weather.

21 December

In Bangalore, rain interrupted the third day of the Test match between England and India.

22 December

The first big freeze of the winter disrupted much of British sport.

Racing went ahead at Ascot only after four stewards' inspections. Tony McCoy might not have felt the cold as he rode five winners. Elsewhere, racing was abandoned due to frost at Newcastle, Uttoxeter and Warwick, the last cancellation putting paid to Cheltenham's novice hurdle which had been transferred there when the Prestbury Park meeting was called off the previous Saturday.

In football, Scotland fared worst with two matches in the First Division frozen off, the entire Second Division and all but one of the Third. All the postponed matches were due to frozen pitches, except at Clyde and Brechin, where the offending factor was snow. Only one fixture could be played in the Highland League. England was not far behind, with five Nationwide League matches called off, all the Division One of the Unibond League and all but three of its Premier Division, as well as most of the Dr Martens and Ryman Premiership games.

In rugby, frozen pitches led to postponement of the Premiership match at Heywood Road (Sale). Down in London Division Four South-West, five minutes were played between Farnham and Richmond before the ref called a halt to proceedings with Farnham's legendary mud being coated with too much frost.

Meanwhile in Bangalore, the Test match between India and England was shortened by rain for the second successive day.

23 December

The bitterly cold weekend continued. In the Rugby Union Zurich Premiership, matches were called off at Kingston Park, Newcastle, at Bristol's Memorial Stadium and at Loftus Road, the last somewhat controversially. Referee Nick Yates turned up at Wasps' ground at 1 p.m. unaware that the game was in danger of not being played. On arrival, Gloucester felt that the pitch was unsafe and that contact in the middle third of the pitch would be dangerous, leaving Yates with no option but to call the match off, less than an hour before the scheduled 3 p.m. kick-off. The travelling Gloucester fans were not amused and club chairman, Tom Walkinshaw, rightly queried what had happened to the undersoil heating and why the pitch had not been covered.

Barnesbury CC's major ten-mile cycling time trial at Seaton Burn, near Newcastle, was cancelled because of snow and ice on the road.

Meanwhile in Bangalore, the final day of the Test match was abandoned, leaving England with the best of the drawn game but also a series loss.

26 December

Once again Boxing Day racecards were interrupted by the winter weather. Meetings at Ayr, Hereford, Huntingdon, Market Rasen, Sedgefield, Towcester and Wetherby were called off as overnight temperatures plummeted. The Ayr executive was criticised for its late abandonment, as some racegoers had already arrived before the programme was called off after a hastily arranged 11.30 a.m. inspection. Kempton Park and Wincanton were run on good going and Wolverhampton on its all-weather track. Not since 1994 has a full schedule of meetings been able to proceed on Boxing Day.

In Scotland, severe frost put paid to Glasgow's rugby union meeting with Edinburgh in the Scottish–Welsh League. A large crowd had been anticipated and the pitch had been covered for four days, but to no avail. David Jordan, the Glasgow chief executive, said that what they really needed was either a heated dome or undersoil heating if they were to play at this time of year. In rugby's other code all but one of the league fixtures in the Northern Ford Premiership were postponed.

Premiership football in England continued uninterrupted but (well) north of the border Aberdeen's match against Dundee was called off. Nationwide League matches were also off at Rotherham, Selhurst Park (Wimbledon), Chesterfield, Carlisle, Hartlepool, Lincoln and York.

28 December

Two climbers roped together fell 200 metres from a blizzard-swept mountain in the Cairngorms when a snow overhang collapsed. They fell vertically for about 30 metres, then bounced and rolled but, fortunately for them, they landed in deep, fresh, soft snow that prevented them falling further to the bottom of a corrie, a steep-sided hollow covered in ice and boulders. Weather conditions were too bad for rescuers to use a helicopter, so the couple were carried down the mountain on stretchers, a journey taking seven hours.

29 December

In the English Football League only northerly Carlisle had their match postponed because of the frozen ground, but in Scotland 12 of the 15 Bell's Scottish League fixtures were postponed, in addition to Motherwell's game against Dundee in the Scottish Premier League.

The frost at Carlisle also put paid to the local race meeting.

30 December

As the winter freeze took its toll on both Scottish Inter-District matches scheduled for the weekend, Alistair McHarg, the former Scottish international and now coach of the Scottish Exiles, suggested that more pitch protection was required. Only the international pitch at Murrayfield has undersoil heating.

31 December

There was a white-out of racing as the continued cold snap claimed all five scheduled meetings, with both Catterick and Fontwell Park called off early on Sunday because the courses were frozen. Cheltenham, Warwick and Wolverhampton followed after early morning inspections on the Monday, the last's all-weather track being unable to cope with an overnight temperature of minus 9°C.

2002
1 January

There were few first foots on the sporting scene. Despite the use of a vast hot-air canopy at Filbert Street, the Leicester v Arsenal match was called off. Referee David Ellery could not guarantee at 7.20 p.m. that the pitch would be safe to play on towards the end of the match, when the cold evening temperatures set in. Although the other Premiership fixtures were OK, only nine Nationwide League games went ahead. Only one Conference fixture survived, Boston v Nuneaton, and that was abandoned after 44 minutes play. Two rugby league friendlies were also off, as well as five of the six scheduled race meetings, the only survivor being the all-weather card at Southwell.

Others welcomed the cold weather. At South Queensferry, 150 enthusiasts took part in the 15th annual 'Loony Dook' charity swim which has become part of Edinburgh's New Year celebrations. And in Holyrood Park another group of revellers spent the first few hours of the New Year racing sleds drawn by huskies. However, a lack of snow forced the racers to use roller-skates rather than blades.

2 January

In Scotland, two Premier League matches were postponed, including Kilmarnock's game against Rangers due to be televised by the BBC, one of the few SPL fixtures not on Sky. Kilmarnock has undersoil heating at its Rugby Park ground and had it switched on for three days, but it simply could not cope with the weather. The

£600,000 system had already failed the previous season when a Scottish Cup tie with Inverness Caledonian Thistle had to be abandoned in the first half because the pitch had become unplayable. The managers of both Rangers and Kilmarnock called for a two-month winter break, similar to those in Germany and Austria, to beat the Scottish weather.

5 January

Ten matches in the third round of the FA Cup and a similar number in the Tennents' Scottish Cup were called off, all but one because of frozen pitches. The one exception was at St Johnstone's McDiarmid Park, where it was too foggy to proceed.

Wolverhampton's all-weather fixture was transferred to Southwell, one of only two of the scheduled race meetings to go ahead. The other was at Lingfield's Polytrack, which was gaining rave reviews.

Elsewhere, the thaw was ruining sport. The fen area at Whittlesey in Cambridgeshire had been deliberately flooded so that a speed skating event could be held on the frozen surface, but it was called off as the ice became unsafe.

The point-to-point season almost did not get under way at Cottenham. The Cambridgeshire course was only declared frost-free after two inspections, and then fog persisted all day, restricting visibility to the home straight, but this was deemed sufficient to allow the meeting to proceed.

7 January

National Hunt racing took place at both Fakenham and Fontwell, the first jump racing since 29 December.

8 January

At the third attempt, the Heineken Cup rugby union match between Newcastle Falcons and Leinster went ahead. Saturday's game had been called off because of a frozen pitch at Kingston Park. The same happened on Sunday, and the fixture was transferred to Headingley, a ground that had subsoil heating. The loss of home advantage did not help Newcastle, who went down 17–15. Newcastle were later fined £13,750 by the European Rugby Union for not using their best endeavours to ensure that their own pitch was playable and not having an alternative venue in reserve.

In India, violent storms in Hyderabad restricted the women's one-day international cricket match to 23 overs a side. England scored 71 for 8 and India knocked the runs off in just 14 overs with the loss of only one wicket.

10 January

Wetherby races survived the frost only to fall victim to thick fog. Two races were run before jockeys, who had trouble finding the fence in the back straight in the novice chase, agreed with the stewards that it was too dangerous to continue.

At the Football Expo in Cannes, David Sheepshanks, chairman of Ipswich, called for a midwinter break in the English football season, but to give players a rest rather than for climatic reasons.

13 January
Greg Rusedski's triumph in the New Zealand Open in Auckland was interrupted several times by thunderstorms and his match finished so late that he was unable to leave New Zealand for the Australian Open until the morning of his first match in Melbourne.

16 January
Heavy overnight rain delayed the start of play until 3 p.m. on the third day of the women's Test at Lucknow between India and England.

17 January
Alex Coomber, a 28-year-old RAF Intelligence Officer, won the skeleton bobsleigh World Cup for the third successive year. The final race was on the Cresta Run's natural track, though most of the other races making up the World Cup series were on man-made tracks.

18 January
Icy roads in the Alps caused Britons Richard Burns and Colin McRae to crash in the season-opening Monte Carlo Rally.

23 January
Several days of heavy rain forced the postponement of the Scottish Premier League match between Dundee and Livingston because Dens Park was waterlogged.

In Ireland, a prolonged downpour caused the abandonment of the Irish Cup fifth-round replay between Glenavon and Lurgan Celtic after 24 minutes. Players, repeatedly slipping on the greasy pitch, found it impossible to make clearances or even pass the ball accurately over any distance. With floodlights unavailable because of ground reconstruction, visibility had deteriorated to dangerous levels by the time the match was called off. Glenavon's manager felt that the game should never have started, a view echoed by the visitors, who lost a player with a fractured knee.

25 January
The Glasgow versus Edinburgh inter-city rugby match fell victim to wintry weather for the second time this season as snow blanketed the Hughenden pitch. It was originally scheduled for Boxing Day, but was postponed because of frost. This time a sudden fall of morning snow put paid to the game. The ground was technically playable but the task of clearing the snow was too huge.

26 January
Three weeks after frost ruined the third round of the FA Cup, rain threatened to

do the same to the fourth round. Matches were called off at Priestfield Stadium and White Hart Lane because of waterlogged pitches. Charlton's game at The Valley against Walsall was saved by 'the whale', the club's water suction machine. Unfortunately, the Premiership team went down 2–1. Almost all matches in the Ryman League, the Dr Martens League and the Unibond League were postponed and, north of the border, another match at Dundee was washed out, this time at Tannadice.

At Bath (appropriately) a waterlogged pitch caused a 24-hour delay in the Heineken Cup rugby union match against Llanelli.

A sodden take-off saw one fence omitted on the steeplechase circuit at Cheltenham, but the abandonment of Lingfield's all-weather meeting after only one race was due to high winds, not deep water.

27 January

Putting covers on their London Road pitch for 48 hours and employing lots of men with forks helped Peterborough's FA Cup match against Newcastle to survive a last-minute inspection.

28 January

Gale-force winds in the worst storms for a decade forced the Scottish rugby union team off the outside pitches to undertake their training session inside the Murrayfield Stadium. Stuart Grimes never even got to training, as he was stuck in a train south of Newcastle for several hours when GNER stopped services because of the high winds.

29 January

Peter Deacon was climbing in the Cairngorms when the area was buffeted by winds of more than 160 kph. He survived by spending two nights in a remote mountain shelter.

Dundee United's fourth-round Tennents' Scottish Cup game against Hamilton fell victim to the weather again. The match was initially scheduled for the previous Saturday but was called off on the morning of the match after heavy rain made the pitch unplayable. The rearranged game was cancelled as the ground was still waterlogged.

30 January

The race meeting at Leicester was abandoned as water from a sewage plant had overflowed onto parts of the track the previous Saturday which, combined with heavy rain, left the course waterlogged.

1 February

During the telecast of the Bradford Bulls' victorious World Club Challenge rugby league match against Australian champions Newcastle Knights, handling errors by the Aussies on a wet and windy evening led the Sky commentator to remark, 'Thank God for the British weather.'

Kelso racing was abandoned after overnight rain added to the 15 mm that had drenched the course the previous afternoon. Hereford was also called off.

2 February

Two matches in the Football League were called off, perhaps to the advantage of Third Division strugglers Swansea and York as it would give them time to regroup. Birmingham v Crewe was abandoned after 58 minutes when the St Andrews pitch was declared unplayable because of waterlogging.

Wetherby races were also waterlogged off.

Competitors in the Volvo Ocean Race were warned to look out for icebergs, while potential members of the British Olympic yachting squad were becalmed in the Miami Olympic Classes regatta.

At Porthcawl, an angler, fishing in almost gale-force conditions, was swept off the pier by a high wave and had to be rescued by the inshore lifeboat.

3 February

World Rally Champion, Richard Burns, gained his first points of the season, despite the second stage of the final day of the Rally of Sweden being cancelled due to a lack of snow.

Welsh rugby fans, many of whom arrived late in Dublin because of ferries and flights delayed by the storms, saw their side humiliated 54–10 by a rampant Irish rugby union team.

In rugby league, Gateshead's game against Swinton was postponed because the pitch was too wet.

4 February

The aftermath of the wet weekend caused a Conference game between Stevenage and Dagenham to be called off because of a waterlogged pitch.

Fitzroy replaced Finisterre as a region on the shipping forecast.

5 February

Carlisle races were called off because of waterlogging.

6 February

Chepstow races were abandoned because of waterlogging on both the chase and hurdle courses. In the 24 hours before the previous afternoon's inspection, 38 mm of rain had fallen.

Pamela Hope was climbing on Aonach Mor near Fort William when she and companion Mike Doyle were struck by an avalanche and swept 60 metres down the mountain. Doyle broke both ankles and was unable to move. Hope dug a snow hole and helped him into a survival bag before setting off for help. During the 12-hour overnight walk she was struck by another avalanche and swept 30 metres, but managed to struggle through the Arctic conditions to alert police at the Nevis Range ski resort.

Snow and strong wind forced the cancellation of the K90 ski-jumping at the Winter Olympics in Salt Lake City where British representative, Glynn Pedersen, was hoping to qualify for the final. He didn't.

Better success came to Kirsty Taylor, who finished second in the Indonesian Open, only her fourth tournament as a professional golfer. The competition was reduced to 36 holes because of a first-day monsoon.

9 February

The Essex Farmers & Union point-to-point, which had already been postponed once, was abandoned, and that at the Vale of Clettwr put off a week because of the impossibly heavy going. Racing at Catterick was also called off.

10 February

The First Division game between Bradford and Birmingham was called off because of waterlogging with standing water in the penalty area. Both managers were consulted by the referee and, after attempting to kick a ball a few times, agreed that the pitch at Valley Parade was unplayable.

11 February

Argentina, in a wet and waterlogged Wales for a hastily-arranged football friendly, could not find a suitable pitch on which to train. Torrential downpours had turned the playing fields of Wales into paddy fields.

12 February

Two Scottish football matches were off because of waterlogged pitches, as was racing at Folkestone and Sedgefield.

18 February

Carlisle racing was off due to waterlogging.

19 February

Torrential rain and flooding wiped out Northern Ireland's football programme, leaving a backlog of 14 matches. At the junior level there was a call for the season to be extended into June.

The Burnley v Birmingham First Division match was postponed because of waterlogging.

22 February

The weather trapped six walkers, who had to be rescued by the Glencoe mountain rescue team after attempting a high-level ridge walk.

23 February

Despite the snow, only one football match was called off – for waterlogging – at Carlisle, who were to have played Cheltenham.

England beat New Zealand in a rain-interrupted one-day match at Napier, thanks to Duckworth/Lewis and man-of-the-match Andrew Flintoff's 4 for 17.

Racing at Haydock was abandoned.

Five rowers were rescued from the Thames after heavy swells overturned their boat.

24 February

The Worthington Cup final, in which Blackburn beat Spurs 2–1, was played under a closed roof at the Millennium Stadium.

Three stranded skiers in the Cairngorms were airlifted out by an RAF helicopter. Less fortunate was cyclist Mark Atkinson, favourite for the Yorkshire Cycling Federation open ten-mile time trial. He was snowbound at home near Fort William.

25 February

Newcastle races were abandoned because of waterlogging.

26 February

The replay of last week's postponed match between Burnley and Birmingham was delayed because the cover that protected the pitch – and enabled the match to be played – proved difficult to remove.

27 February

There was no National Hunt racing for the second day running as heavy rain saw off both Chepstow and Wetherby.

28 February

Perspiring in the high humidity and 35°C temperatures, British duo Barry Lane and Alastair Forsyth both scored an eight-under-par 63 course record in the Carlsberg Malaysian Open at the Royal Selangor Club in Kuala Lumpur. Both looked as if they had been for a swim rather than a round of golf! Forsyth went on to win his first European tour event.

1 March

On a bright sunny day, Kelso racing was off because of earlier snow and rain. This was the third consecutive meeting to be called off at the Borders course.

2 March

Setting out to assist a climber who had fallen more than 60 metres down the north face of Ben Nevis, the Lochaber mountain rescue team was forced to turn back by bad weather. They finally reached the stricken mountaineer, John Hunston, at 4 a.m. the following day.

9 March

Two climbers, Ian Collier from Alloa and Douglas McQuaker from Glasgow, were killed by an avalanche on Beinn Dearg in Braemore, Wester Ross. Another two climbers became disoriented in the appalling weather on Ben Nevis and were rescued after spending a night on the mountain.

Snow also caused the postponement of Darlington's Third Division match against Plymouth, but only after the south-coast team had travelled north.

The Harkaway Club's point-to-point fixture fell victim to high winds.

10 March

Westcoast Western lost the opportunity to open up a seven-point gap at the top of the Spektra Systems women's hockey league when high winds blew over the goalposts at Bellahouston. This forced the abandonment of the game against Western A with the Westcoast side one goal up.

13 March

Newton Abbot races were called off because of waterlogging.

14 March

The harsh sunlight in Doha gave Colin Montgomerie a painful headache during the first round of the Qatar Masters, but he still shot a four-under-par 68.

16 March

The ground was so heavy at Uttoxeter that the Midlands Grand National was won by The Bunny Boiler in a time over one and a half minutes slower than standard for the race. At least they raced, whereas both the Cheshire point-to-point and the meeting at Newcastle were abandoned due to waterlogging.

17 March

The Lochaber mountain rescue team were called to a walker who fell over 100 metres after slipping on snow on Stob Coire Easain at Glen Spean near Fort William.

19 March

Racing at Exeter was cancelled.

23 March

Ronnie Moore, manager of Rotherham United, ordered the soggy, lumpy pitch at Millmoor not to be rolled so as to negate the skills of Division One leaders, Manchester City. The ploy worked and relegation-threatened Rotherham secured a vital point with a 1–1 draw.

In Dublin, a gale that made it almost impossible to stand up straight forced Paula Radcliffe to revise her plans to lead from the start in the World Cross-

Country Championship. She still won by 20 metres to become the first woman to retain the title in a decade.

27 March
Fears that wet weather could aggravate the effects of one of the highest tides of the year led to press speculation that the Boat Race might be postponed for 24 hours, or even rowed in the reverse direction. Neither happened and Oxford gained only their third victory in the last ten years.

29 March
Two climbers plunged to their deaths on Stob Coire nan Lochan in Glencoe. The weather had been good but closed in with snow, wind and mist. In the poor visibility the party strayed off their route.

1 April
A bank holiday fishing trip off the Wirral coast that began in fine weather almost ended in tragedy when a sudden squall capsized the anglers' boat. Wirral's chief lifeguard urged anglers to get thorough weather reports before setting off.

5 April
Heavy rains two weeks before the Scottish Nomad river fishing championships had given the River Tummel, near Pitlochry, a peaty look, ideal for competitive angling. However, the water temperature was still low, which made the fish lethargic, and only six were caught.

6 April
A crowd of 63,500 attended the Martell Grand National, a modern-day record. Charles Barnett, managing director of Aintree racecourse, attributed this partly to magnificent weather. The BBC blamed the weather for a decline in the number of those watching the event on television.

7 April
The Algarve Portuguese Open at Vale do Lobo was cut to only two rounds because of heavy rain and blustery wind. Leaders Carl Pettersson and David Gilford arrived at the course at 10 a.m., waited for six hours and then were sent out on a sudden death play-off at the 18th hole, which was won by the Swede.

8 April
Scarborough, desperate to avoid relegation from the Football Conference, had the luck of the weather in their encounter with Leigh RMI. A wind change at half-time meant that the Yorkshire side had the benefit of the breeze (later almost a gale) throughout the game. They secured a valuable point in a 1–1 draw.

9 April

Coventry road cycling club enjoyed the best weather in memory for its first evening time trial of the season, in stark contrast to the previous year when the event was washed out.

12 April

The warm spring weather allowed cricket groundstaff to use their mowers and rollers to produce good pitches for the beginning of the first-class season. Worcestershire's director of cricket, Australian Tom Moody, said it was the best weather at the start of the season that he had experienced in the 12 years he had spent in England. He added, 'It's *almost* been nice enough to train without a jumper on.'

13 April

Dry weather caused problems for anglers who fish the waters around the Lough Neagh in Ireland. The level of the rivers has been lowered and fish cannot get upstream.

15 April

The threadbare turf at Hampden Park was ripped up, to be replaced with grass grown in an area near the Dutch–German border with a winter climate similar to that of Scotland.

19 April

Ayr racecourse was watered to soften the going for the next day's Scottish Grand National.

In the Rally of Cyprus it was so hot that world champion, Richard Burns, had to have oxygen treatment after one of the stages.

21 April

The Seve trophy match at Druid's Lodge, Dublin between golfers from continental Europe and those from Great Britain and Ireland had its third day of wet and windy conditions.

In the Rally of Cyprus, Colin McRae, the overnight leader, went out with dry weather tyres but it rained and he rolled his car on two of the stages and slipped to seventh place.

Rain prevented play before lunch on the third day of the match between Durham and Middlesex at the Riverside. It could not save the home side, who lost by ten wickets.

22 April

It was announced by the FA that the England World Cup squad would wear revolutionary frozen training jackets in a bid to keep cool in the intense heat and humidity of Japan. The sleeveless white jackets, developed by NASA, are designed to help athletes sweat less and avoid dehydration.

Tim Packwood, the new head groundsman at New Road, Worcester was praised by both teams after the game against Gloucestershire. He had to contend with three floods during the winter, but a dry spring and the installation of new drainage enabled him to produce a pitch that had sufficient pace to encourage the quick bowlers while allowing batsmen to play their shots.

27 April

Trainer James Fanshawe walked the track at Leicester to check that the recent dry spell had not made the ground too firm for his horse, Warningfold, who went on to win the Group Three Leicestershire Stakes for the third time.

The cricket season returned to normality as rain prevented play at Canterbury, Derby and Chelmsford and disrupted it at Lord's, Southampton, Hove and Headingley. In Scotland, the first of the new super-league games between Thistles and Saltires was abandoned after only 18 overs because of rain in Edinburgh.

28 April

The first round of the Benson & Hedges Cup saw Yorkshire, Essex and Warwickshire all win courtesy of Duckworth/Lewis, and the match between Derbyshire and Lancashire was abandoned as a draw after only 4.3 overs were possible.

Paul McGinley won the Smurfit PGA Irish Region Championship at Westport without hitting a ball in the last round. The worst weather he had experienced in any part of the world caused its abandonment.

30 April

Benson & Hedges matches at both Taunton and Lord's were abandoned without a ball being bowled due to rain.

Appendix

Temperature Conversion Chart

Celsius/Fahrenheit

C	F	C	F	C	F	C	F	C	F	C	F
−50	−58.0	−33	−27.4	−16	3.2	1	33.8	18	64.4	35	95.0
−49	−56.2	−32	−25.6	−15	5.0	2	35.6	19	66.2	36	96.8
−48	−54.4	−31	−23.8	−14	6.8	3	37.4	20	68.0	37	98.6
−47	−52.6	−30	−22.0	−13	8.6	4	39.2	21	69.8	38	100.4
−46	−50.8	−29	−20.2	−12	10.4	5	41.0	22	71.6	39	102.2
−45	−49.0	−28	−18.4	−11	12.2	6	42.8	23	73.4	40	104.0
−44	−47.2	−27	−16.6	−10	14.0	7	44.6	24	75.2	41	105.8
−43	−45.4	−26	−14.8	−9	15.8	8	46.4	25	77.0	42	107.6
−42	−43.6	−25	−13.0	−8	17.6	9	48.2	26	78.8	43	109.4
−41	−41.8	−24	−11.2	−7	19.4	10	50.0	27	80.6	44	111.2
−40	−40.0	−23	−9.4	−6	21.2	11	51.8	28	82.4	45	113.0
−39	−38.2	−22	−7.6	−5	23.0	12	53.6	29	84.2	46	114.8
−38	−36.4	−21	−5.8	−4	24.8	13	55.4	30	86.0	47	116.6
−37	−34.6	−20	−4.0	−3	26.6	14	57.2	31	87.8	48	118.4
−36	−32.8	−19	−2.2	−2	28.4	15	59.0	32	89.6	49	120.2
−35	−31.0	−18	−0.4	−1	30.2	16	60.8	33	91.4	50	122.0
−34	−29.9	−17	1.4	−0	32.0	17	62.6	34	93.2		

Select Bibliography

Bale, John, *Landscapes of Modern Sport* (Leicester UP 1994)

Barrett, Norman, *Daily Telegraph Football Chronicle* (Carlton 1994)

Benedictus, David, *Sunny Intervals and Showers* (Weidenfeld & Nicolson 1992)

Bird, Dickie, *My Autobiography* (Hodder & Stoughton 1997)

Blakemore, Colin & Jennett, Sheila, *The Oxford Companion to the Body* (Oxford UP 2002)

Bond, Bob et al., *The Complete Book of Sailing* (Chancellor 1997)

Burroughs, William James. *Does the Weather Really Matter?* (Cambridge UP 1997)

Cashmore, Ellis, *Sports Culture* (Routledge 2000)

Dheensaw, Cleve, *The Commonwealth Games* (Queen Anne Press 1994)

Eden, Philip, *Weatherwise* (Macmillan 1995)

Evans, R.D.C., *Bowling Greens* (Sports Turf Research Institute 1992)

Evans, R.D.C., *Winter Games Pitches* (Sports Turf Research Institute 1994)

Frindall, Bill (ed), *The Wisden Book of Test Cricket*, Vols 1 and 2 (Macdonald Queen Anne Press 1990)

Greenberg, Stan, *Whitaker's Olympic Almanack* (The Stationery Office 2000)

Holford, Ingrid, *The Guinness Book of Weather Facts and Feats* (Guinness 1982)

Kent, Michael, *The Oxford Dictionary of Sports Science and Medicine* (OUP 1994)

Martin-Jenkins, Christopher (ed), *Seasons Past – the Cricketer Diaries* (Stanley Paul 1986)

Murray, Francis, *The Open: A Twentieth Century History* (Pavilion 2000)

Nawrat, C., Hutchings, S. & Struthers, G., *The Sunday Times Illustrated History of Twentieth Century Sport* (Hamlyn 1995)

Parke, Hilary, *Ski & Snowboard Scotland* (Luath Press 1998)

Rousmaniere, John, *Fastnet Force 10* (Nautical 1980)

Samuel, John, *The Guardian History of the Open Golf Championship* (Fourth Estate 1992)

Sly, Debby, *Olympic Eventing Masterclass* (David & Charles 1996)

Sly, Debby, *Badminton Horse Trials* (David & Charles 1999)

Stirling, Robin, *The Weather of Britain* (dlm 1997)

Thomas, Malcolm, *Weather for Hillwalkers and Climbers* (Alan Sutton 1995)

Unsworth, Walt, *Encyclopaedia of Mountaineering* (Robert Hale 1975)

Ward, Andrew, *Soccer's Strangest Matches* (Robson 1992)

Welch, David (ed), *The Daily Telegraph Century of Sport* (Macmillan 1998)